VETERAN TREES
INITIATIVE

Foreword

As we approach the new millennium spare a thought for the real veterans of this one - trees several hundred, perhaps one or two even a thousand, years old. This book, produced by the Veteran Trees Initiative, is the first dedicated solely to the care of the oldest living things in our countryside.

Veteran trees are a significant part of our historic, cultural and ecological heritage, treasured by many generations; in the past because of their economic and social value or as elements of picturesque or romantic landscapes and more recently as a result of our increased understanding of their considerable ecological importance.

In parallel with this has come greater understanding of the management they need. Many of the techniques familiar to our forebears have been lost or swept aside and current techniques have been re-assessed, notably through the lively meetings and discussions within the Ancient Tree Forum. The Veteran Trees Initiative is therefore issuing this guidance as our current understanding of best practice, although accepting that it may need to be revised in the future.

The handbook gives practical advice on all aspects of veteran tree management, from the importance of sometimes doing nothing at all to taking positive action for individual trees, their habitats and dependent species. All is set in context by an understanding of the way in which trees grow, age and decay.

This handbook is one of the most significant publications of the partnership which is the Veteran Trees Initiative. I am sure it will become an invaluable reference handbook for all those, from arboricultural professionals to countryside advisors, involved in the conservation and management of veteran trees whether for their ecological, landscape or historic value. Other elements of the work of the Veteran Trees Initiative are now reaching fruition and will be published shortly. Together they make a major contribution to the implementation of the Lowland Wood-pasture and Parkland Habitat Action Plan, which English Nature leads.

Baroness Young of Old Scone
Chairman, English Nature
August 1999

Acknowledgements

This book is the result of a great deal of work by very many people. At several stages a draft was sent out to many different people, from a range of disciplines, who took the time to make copious comments. The various comments and suggestions improved the final version substantially and contributed greatly to the contents. I would also like to thank all the members of the Ancient Tree Forum who have attended the field meetings and helped to improve our knowledge of veteran trees.

I am especially grateful to David Lonsdale for contributing greatly to the physiological aspects of tree growth and decay, Neil Sanderson for helping with the section on lichens and mosses, Sheila Wright for help with bats, Maurice Waterhouse and the RSPB for the log-pile design and Neville Fay and Mark Frater for all-round support and encouragement. Of the many that read through the script at different stages special thanks are due to Rob Green for such comprehensive comments also Tony Robinson, Rachel Thomas, Keith Kirby, Alan Richardson and Roger Key for spending so much time reading through the various drafts. (Thank you Roger for those comments that brought some light relief!)

I am grateful to all those who made comments on the various drafts of the book, helped with specific sections and those who helped produce the glossary; my apologies to anyone I may have inadvertently omitted:

Martyn Ainsworth	Jeanette Hall	Francis Rose
Keith Alexander	Paul Harding	Neil Sanderson
Jill Butler	Peter Holmes	Pete Shepherd
Alan Cathersides	Harriet Jordan	Paul Sinaduri
Fred Currie	Richard Jefferson	John Smith
Jeremy Dagley	Roger Key	Helen Stace
Caroline Davis	Keith Kirby	Rachel Thomas
Lesley Davies	Norman Lewis	Tom Wall
Neville Fay	David Lonsdale	Bob Warnock
Roy Finch	Mike Martin	Ashley Wheal
Vikki Forbes	David Maylam	John White
Adrian Fowles	Peter Quelch	The Whiting family
Mark Frater	Oliver Rackham	Ray Woods
Rob Green	Paul Read	Sheila Wright
Ted Green	Alan Richardson	
Jackie Haines	Tony Robinson	

The illustrations were drawn by Sarah Wroot. Several were adapted from originals by Neville Fay and one (figure 11) from an original by Alex Shingo. I would like to thank the following for permitting the use of their photographs for the following figures:

Keith Alexander (7, 9, 25, 26, 33)
The Corporation of London (24a, 44a, 51)
English Nature (1, 13, 24)
Roger Key (43, 54)
Jen Read (21)

Rachel Thomas (48)
English Heritage (4, 31, 32)
Ted Green (14, 15, 38)
The National Trust (22, 45, 46, 50)

Helen Read (February 2000)

Veteran Trees | A Guide to Good Management

Contents

Preface

Appendices

List of figures

Preface

Veteran trees are a precious part of our heritage but they need care and attention in order to survive into the future. This book is intended for those who manage this type of tree and those who give advice on them.

It is not possible to give a description of tried and trusted methods for managing old trees. The methods used by our ancestors were not well recorded, and in addition we have a problem which they largely did not have; that of restoring working trees which have not been managed for many years. The intention here is to make suggestions on methods likely to be successful based on the experiences of many people and a review of the available literature. It represents best practice at this time. There are no specific recipes that can be followed; instead this book offers you the best available evidence so that you can make an informed judgement for your particular circumstances. Recent management practices are outlined, those that have worked and those that have not. It is important to realise that because a technique has worked well on one individual or group of trees, because of their unique responses to environmental conditions it does not necessarily mean that it will work on others. However it is usually worth considering.

The trees we are working with now have seen many changes in their long lives. On most sites (most, because there are exceptions) a few more months of careful deliberation over what to do are wisely spent and are just a moment in the life of these organisms.

Perhaps the over-riding principle of ancient tree management is not to do the same thing to every tree on a site at the same time. If you are sure that what you are intending to do will have the desired effect it is still better to spread the work over several years. Factors such as the weather cannot be planned for. Work in small stages. The longer ago that the tree was last worked on, the smaller the amount of tree surgery that should be done at one time.

This is a reference handbook, to read and to check with, in advance of making decisions on the practical aspects of the management of veteran trees, rather than a book to read from beginning to end. It will help you to decide whether it is appropriate to carry out management work on veteran trees and, if so, which points should be taken account of in terms of timing and working method. It explains how trees grow, age and decay and what impact this has on their wildlife and landscape value. It discusses the management of land around veteran trees, whether for nature conservation or cultural reasons as well as the management of individual trees where the conservation of particular species is of concern.

The first few sections introduce veteran trees, outlines their importance and why they may need managing. It may be possible to skip some of this background material if you already have experience of these topics although you should check both the general and specific sections when seeking information on specific aspects of veteran tree management. Consequently there may be some repetition but less danger of missing important areas. Finding your way around should be easy as many paragraphs are cross-referenced and the index at the back includes both individual species and topics which many readers will be interested in. I recommend that you become familiar with the content and scope of the book early on before using it in real decision making.

Throughout the text, boxes are used to give more detailed information. Further reading is suggested at the end of each chapter with the full references included at the end of the handbook. The legal aspects of veteran trees are a specialist topic and is addressed in a separate Veteran Trees Initiative publication. Throughout the handbook, English names for organisms are given where possible. Appendix 1 lists the scientific names for those trees mentioned in the text while other appendices present further details and sources of information on veteran tree management.

The main focus of this book is trees in lowland situations although examples from upland areas are given where possible. However, most of the recent experience with veteran tree management has come from lowland Britain.

There is plenty of scope for careful innovation in the field of ancient tree management by experienced workers. Try different methods to see which works but ensure that you record what you do and learn from the results. Tell other people about your successes and failures. Do not though threaten your precious trees by testing theories for which the answers are well known or taking unnecessary risks. For example, it is worth while experimenting with different methods of cutting, but cutting during bud burst is certain to be damaging.

Take care of your trees. Remember that they are an invaluable asset and an important part of the nation's heritage; above all, enjoy them.

Helen Read
1999.

Disclaimer

The publishers, author and contributors will not accept any liability for loss or damage that may be suffered by any person as a result of the use, in any way, of the information presented here.

VETERAN TREES

Chapter 1   Introduction

Veteran trees are an integral and valuable part of the lowland British landscape. They are the old trees in woodland and parkland, the gnarled oaks in the hedgerows and the decaying pollard willows along riversides. Our ancestors valued these trees as vital assets; they were part of their subsistence and economy as well as objects of religious and social interest. In our more urban society most veteran trees are no longer retained and managed for their produce. A few have become tourist attractions because of their historical connections, but most are forgotten and neglected. Many more have already been lost, felled to make room for development, intensive agriculture and forestry, or for safety reasons.

Recently there has been a resurgence of interest in these elderly trees. Their biological, historical and cultural importance is slowly being recognised together with their aesthetic appeal and the unique contribution they make to the landscape. These trees are as much a part of our heritage as stately homes, cathedrals and works of art, and are a favoured subject of many of our most important paintings and engravings. Many veteran trees, such as pollards, exist as a result of man's handiwork.

Britain is of European importance for the large number of old broadleaved trees still surviving here. This is largely a result of historical factors which have allowed veteran trees in Britain to survive while in other European countries, old trees have severely declined in number or were rarely allowed to remain.

Across Britain the distribution of veteran trees is patchy. In some parts of the country they are very scarce, in other areas they may be surprisingly abundant. Once you have begun to notice old trees you start seeing them everywhere, village greens, churchyards, open farmland and urban streets.

Interest in ancient trees has been stimulated in recent years by the formation of the Ancient Tree Forum (ATF), which originated as a discussion group concerned with their management. Two meetings on veteran tree management organised by the Corporation of London resulted in publications (Read 1991, 1996) that have been widely disseminated. Then in 1996 the Veteran Trees Initiative (VTI), a partnership started by English Nature, was launched with the aim of promoting the conservation of veteran trees wherever they occur. Publications arising from the Initiative so far include an introductory leaflet (English Nature 1996), a conference report (Bullock & Alexander 1998) and a book about Moccas Park (Harding & Wall in press.). The interest of the general public has been stimulated by various television programmes (eg *Meetings with Remarkable Trees* and *Spirit of the trees*), Trees of Time and Place, the Great Trees of London project and the WATCH tree pack for children. The conservation of veteran trees is being further encouraged through the Lowland Wood-pasture and Parkland Habitat Action Plan (see Appendix 2) and promotion by the Veteran Trees Initiative of the inclusion of old trees in Local Biodiversity Action Plans. The increasing number of regional surveys of veteran trees will also help to establish their distribution and abundance.

Chapter 2 What are Veteran Trees and why are they important?

2.1 What is a veteran tree?

- The term veteran tree is one that is not capable of precise definition but it encompasses trees defined by three guiding principles:
  - trees of interest biologically, aesthetically or culturally because of their age;
  - trees in the ancient stage of their life;
  - trees that are old relative to others of the same species.

- The girth of a tree is not a reliable criterion because different species and individuals of tree have very different life spans and grow at different rates.

- Veteran trees can be identified by the presence of specific characteristic as listed in the main text.

A veteran tree can be defined as: **'a tree that is of interest biologically, culturally or aesthetically because of its age, size or condition'**. Some trees are instantly recognisable as veterans but many are less obvious.

An alternative approach used by some people is to consider that the veteran, or ancient, stage is the final one in the life of a tree when the cross-sectional areas of successive annual rings in the main stem begin to decrease progressively. (Before this stage, successive rings will have already narrowed, but their areas will have been roughly constant, owing to their increasing girth.) In turn, the amount of leaf area that can be supported by the reduced annual increment eventually results in dieback of the crown. For this reason veteran trees are rarely tall with large crowns. In theory this definition sounds fine but in reality this growth phase may not be clearly recognisable even though it may be the longest one in the tree's life. Dryden describing oaks is reputed to have said, 'three centuries he grows and three he stays, supreme in state, and in three more decays'.

Size alone is a poor characteristic for determining veteran status, although some rules of thumb exist (see box). Different species of tree may grow to very different maximum sizes. The simple comparison of a huge mature oak tree (Figure 1) with a small gnarled veteran hawthorn (Figure 2) illustrates this point. In addition the same species can grow to very different sizes in different situations and conditions.

Figure 1. *See colour plate page 81.*
Figure 2. *See colour plate page 81.*

A rough rule of thumb can be adopted for species, eg oak, in relation to size:

- Trees with a diameter at breast height of more than 1.0 m (girth 3.2 m) are potentially interesting.

- Trees with a diameter of more than 1.5 m (girth 4.7 m) are valuable in terms of conservation.

- Trees with a diameter of more than 2.0 m (girth 6.25 m) are truly ancient.

Absolute age is also a poor indicator of ancient status for trees. Different species tend to live for varying numbers of years; thus age can only be used when considered in comparison with other trees of the same species. At 100 years of age a birch would be old and a willow extremely old. At 200 a beech would just be starting to become interesting, an oak just maturing and a yew still beginning. One age-related definition sometimes used is that of an individual older than about half the natural life span for that species (but defining the natural life span is also a challenge!)

The increasing complexity of the tree with age results in a range of features in root, trunk and branch; these features are often good indicators of old age.

2.1.1 Characteristic features found on veteran trees

Listed below are characteristic features of veteran trees (see also Figure 3). The more the tree has, the stronger the indication that it is a veteran:

- Girth large for the tree species concerned
- Major trunk cavities or progressive hollowing
- Naturally forming water pools
- Decay holes
- Physical damage to trunk
- Bark loss
- Large quantity of dead wood in the canopy
- Sap runs
- Crevices in the bark, under branches or on the root plate sheltered from direct rainfall
- Fungal fruiting bodies (e.g. from heart rotting species)
- High number of interdependent wildlife species
- Epiphytic plants
- An 'old' look
- High aesthetic interest

Figure 3★. *Diagram to show the features characteristic of a veteran tree.*

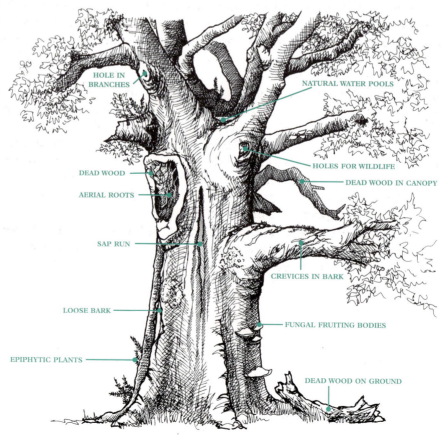

In addition the tree may also:
- Have a pollard form or show indications of past management
- Have a cultural/historic value
- Be in a prominent position in the landscape

One of the difficulties of using the above list as an indicator of veteran status is that young且 trees which have been physically damaged; eg by fire, can show these features, whereas some veterans may exhibit very few.

While a veteran tree is alive, part of its value for wildlife is that it is a self-renewing resource. However, dead veteran trees continue to be valuable for wildlife because of their slow rate of decay. Such trees are often greatly under-valued but they should be treated with almost as much respect as living ancient trees.

It is not surprising that different people or organisations have slightly different ideas about how to recognise veteran trees. (Several surveys have used size as the sole, or major, criterion but this can be misleading, as demonstrated above.) On the whole it is best to err on the side of caution and remember that, even though your tree might not quite be a veteran yet, with care it will become so, helping to ensure the continuity of ancient tree habitats.

How have veteran trees survived in Britain?

A variety of factors have helped some veteran trees to persist in the British countryside (although there have been huge losses in the past):

- Many (if not most) trees were 'working trees' (Green 1994), ie their wood and leaves were used, as an essential part of everyday life, for much of the recorded history of Britain. The management methods that predominated may have helped individual trees to survive (eg pollarding and coppicing).
- Very large trees were time consuming to fell and if the tree was rotten inside then the timber value was considerably reduced. This helped the survival of, for example the oaks at Birklands in Nottinghamshire, and Windsor Forest.
- Continuity of ownership on many estates. Changes of ownership may lead to decisions to remove old trees.
- Common land rights. Many veteran pollards were situated on common land and were owned by one person while others managed them (unlike coppice woodlands). This situation helped perpetuate their survival.
- Veteran trees have been incorporated into successive changes in the landscape. This occurred prior to and during the enclosure of land to form parkland and during the creation of design landscape parks by Lancelot (Capability) Brown and Humphry Repton. The retention of veteran trees was considered to give the parks an air of respectable antiquity (Rackham 1991).
- In Britain veteran trees have generally been revered and respected. Some old trees and some species of tree are regarded as sacred, which helped in their survival (eg yew trees in churchyards).

2.2 Why are veteran trees important?

..those grey old men of Moccas, those grey, gnarled, low-browed, knock-kneed, bowed, bent, huge, strange, long-armed, deformed, hunchbacked misshapen oak men that stand awaiting and watching century after century biding God's time with both feet in the grave and yet tiring down and seeing out generation after generation.
The Reverend Francis Kilvert, 1876.

The 'grey old men' from all over Britain have inspired artists and writers throughout the centuries. They have featured in the paintings of Thomas Hearne and John Peddar while Thomas Gray thought of them as *'Reverend vegetables .. always dreaming out their old stories to the wind'* (in a letter to Horace Walpole 1737). Notable specimens have been revered in the past and some have become tourist attractions, the Major Oak at Sherwood being a famous example. Through their individuality, veteran trees have high intrinsic appeal and are strong 'characters'. A few people have found the distorted shapes of old, repeatedly cut, pollards grotesque but even then they still managed to inspire, as seen in the comments of a journalist writing about Epping Forest. *'Short, shabby, scrubby, indescribably mean and ugly they were - something like warty railway sleepers with a shock head of twigs'* (journalist in Cole 1894).

All veteran trees are of historic interest; each is a survivor from the past, a relict of a former landscape. In addition to their importance as natural habitats, they are a valuable part of our cultural heritage. This historic interest lies both in the individual trees, each of which is a living document telling us of past management practices and ways of life, and in the tree's place in the wider countryside. The distribution of veteran trees in the landscape gives an insight into former land use patterns. Each tree, or group of trees, deserves individual consideration and study, perhaps combined with document-based research in local and national archives to enable us to understand fully its historic context and importance.

Within the existing agricultural landscape, veteran trees are most frequently found as markers along old boundary banks and occur in long established hedgerows. As such they contribute to our knowledge of historic land divisions. Other veterans, particularly pollarded willows, chart the course of rivers, or other water channels such as mill-leats, now often dry, visible on the ground only as archaeological traces. Veteran trees are closely associated with wooded commons, now almost lost as a form of land management, and are also frequent in churchyards where their existence and location can be of great significance, sometimes pre-dating Christianity.

Some of the highest concentrations of veteran trees are found in current and former parkland. However early in date the origins of a park, all parks were developed from an existing landscape and in most cases features of the pre-park landscape were incorporated into the new enclosure. Parks as they are today are invariably the result of several changes in ownership and fashion with each phase leaving its mark on the landscape. Successive designs tended to incorporate valuable features that were already present and veteran trees were often considered to add maturity (see Figure 4). Our generation is not the first to treasure their presence.

Figure 4. *See colour plate page 82.*

Some oaks alive today can be traced back to the medieval period or, in the case of some yew trees, before the start of Christianity. Many more trees predate great architectural structures of the 17th and 18th centuries that we admire and conserve.

Despite surviving centuries, they are now largely at our mercy through the rapid pace of change brought about by modern technology. It only takes a few minutes to condemn a tree that has lived through more changes in its time than we can ever imagine.

In brief, veteran trees are of importance because:

- **They have aesthetic appeal and cause inspiration.**
- **They may have a particular historic link, ie be associated with a specific person or event.**
- **They often illustrate past land use or cultural landscapes. For example veteran trees are often found on wooded commons, in parkland, as boundary or field markers and in ancient farmland landscapes.**

VETERAN TREES
INITIATIVE

- They may be part of a designed landscape or garden. Many formally laid out gardens contain veteran trees, and avenues may be comprised of them. They are also especially abundant in some of the landscapes designed by Lancelot Brown, Humphry Repton and other landscape architects of the 18th and 19th centuries.
- They are especially important for biological reasons, providing conditions suitable for a wide range of other plants and animals, many of which require the very special environment created in an old tree. They have been likened to a block of flats.

> *Fungal rotting of the heartwood and dead limbs results in a diversity of micro-habitats suitable for other organisms including a potentially very wide range of invertebrates, dependent on such different micro-habitats, and birds such as woodpeckers which prey on them. Epiphytes such as mosses and lichens may require the old bark characteristic of veteran trees to grow on. Although some of the organisms are generalists, many are extremely specialist and are confined to veteran trees. Old trees, as a consequence of their rarity, harbour large numbers of rare and threatened species. The biological importance of a tree is greater if it lives long enough to perpetuate the continuity of habitats for future generations.*

- They provide an air of stability in an ever changing world.
- Very old trees are more likely than younger trees to be descendants of the trees of the natural wildwood that colonised Britain after the last ice age. This makes them a reserve of important genetic material. (However, some veteran trees have been demonstrated to be of introduced origin.)
- They may be an important gene pool of trees showing particular characteristics, eg disease resistance or good epicormic growth (beneficial for good growth after pollarding but not for good quality commercial timber).
- The annual rings of old trees are historical records in their own right. They illustrate past climate changes or cutting treatments, and the chemical nature of the wood is a potential resource for research into past climates, pollution levels etc. (However, the decay process removes the rings as the tree becomes hollow).

In addition, Britain has one of the highest populations of veteran trees in Europe (along with Greece and Spain).

> ### Why populations of veteran trees are more important than isolated trees
>
> - The more trees, the more alternative niches there are.
> - Organisms that require precise micro-habitats are more likely to find enough to support viable populations.
> - Groups of veteran trees can yield more information about past practice, and their population structure than single trees.
> - Groups of veteran trees are less threatened by change than single trees.

2.3 Types and locations of veteran trees

Veteran trees found in Britain today can be described and assigned to categories according to their origin and past management. There are three widely found types of veteran tree: maidens, coppice and pollards.

2.3.1 Maiden trees

These are trees that have a trunk extending from the base to the upper crown and have not been cut in any way other than perhaps minor tree surgery. They may be woodland trees that have grown up with other trees close by and thus tend to have a 'narrow' profile with a tall stem and small canopy, or they may be open grown with a much wider crown and bigger branches lower down the trunk. Open grown trees may subsequently be surrounded by younger woodland and a woodland grown tree may be exposed by the felling of surrounding woodland. Trees planted and left to grow without intervention, for example as part of an avenue or designed landscape, are usually maidens.

2.3.2 Coppice stools

A coppiced tree is one cut near ground level, then allowed to produce new shoots from the stool. The shoots from a block of woodland are cut repeatedly in cycles of varying length depending on the size of sticks or poles required. A range of tree species produce coppice growth, some much more readily than others. Although the growth from a coppice stool is usually quite young, the stool itself can be extremely old. These veteran trees can be very different in shape to maidens or pollards. Generally speaking, the larger the stool width or height, the older the stool (within species). For example, an ash stool 2 m in diameter has been estimated to be over 500 years old and a 16 m diameter lime stool was estimated at 2000 years old (John White pers. comm.). Very old coppice stools may rot out in the centre, leaving a circle of apparently younger stools. Coppice largely occurs in woodlands managed specifically in this way but is also found on ancient wood or boundary banks, along rivers and in hedges. Some coppice has not been cut for many years and may take on a tree-like appearance.

2.3.3 Pollards

A pollard is a tree cut like a coppice but well above the ground (Figure 5). Usually, the reason for cutting high up was to allow animals to graze among the trees without damaging the next crop of branches by browsing them. Thus the height of the pollard was partly determined by the type of animal (that to deter sheep did not need to be as high as that where cattle were grazed). The products of pollarding were leaves, twigs and bark for animal fodder, bark for tanning and wood for fuel and charcoal. Pollards were probably first cut when the maiden tree was quite young and small in girth with subsequent cuts made at regular or irregular intervals. In some places it seems unlikely that pollarding was carried out in such a formal and regular cycle as occurred with coppicing. The proportion of branchwood removed at each cut was probably also variable. In the case of willow trees all the growth was removed each time. For other trees, e.g. beech, it seems likely that some branches were cut while others were left (see also chapter 4). Pollarding is sometimes taken to be the total beheading of the tree, but here a pollard is taken to mean a tree cut back once or more (to a similar point) by removing a substantial number of branches. The presence of a number of pollarded trees in a group is often a good indication that the area was wood-pasture or parkland at some point in the past, though willow pollards are not necessarily associated with grazed systems.

Many, but by no means all, of the veteran trees in Britain today have been pollarded at some stage in their life. However, pollards are not necessarily old trees. The majority of young trees pollarded in recent years can be found in urban situations and on some sites with veteran pollards where owners/managers are now starting to create new pollards.

Figure 5. *See colour plate page 82.*

2.3.3.1 Where are veteran pollards found?

- Wooded commons where commoners had rights to graze animals and cut or collect wood (eg Ashtead Common, Surrey). Village greens are another form of common where veteran pollards may be found.
- Parklands. Private, enclosed, land usually grazed by deer, occasionally cattle (eg Moccas, Herefordshire).
- Wooded Royal Forests. Land governed by special laws where deer owned by the Crown or a wealthy land-owner were kept. Often these forests incorporated existing areas with commoners' rights (eg Hatfield Forest, Essex)
- As farm trees for a local wood supply, scattered about the farm and sometimes in hedges. Some farmsteads almost had miniature 'parks' around them (eg in the Lake District). In places this can produce a pollard landscape (eg parts of the Cotswolds).
- Upland grazed woodlands. Most upland woods were unenclosed and grazed by sheep, cattle or deer especially during the winter months. This practice continues today, although in the 18th and 19th centuries many woods were enclosed. The trees in these woods were sometimes pollarded, especially oak and ash. Some upland woods were summer grazed and may have contained a wider range of pollarded species, including alder, hazel, birch, ash and rowan.
- As boundary markers between, for example, parishes or areas of different ownership (eg East Anglia and Kent). Also as boundaries between different 'panels' of woodland, where at least some were grown out from layered hedges (D. Maylam pers. comm.).
- As elements within the designed landscape (eg clumps, avenues and pleasure gardens).
- Churchyards. Although not pollarded, yew trees are associated with churches, other species such as lime were frequently planted and may have been pollarded to keep them manageable. In addition, many churchyards were grazed in the past.
- Beside rivers and in withy beds (for the production of willow branches for baskets, etc.) Often these trees are in grazed meadows but sometimes they are low pollards, cut above ground for the ease of cutting rather than to protect shoots from grazing animals. Black poplars were pollarded in damp meadows close to rivers. Pollards were also used to help stabilise banks along roadsides in areas of wet fen and bog.
- As urban or street trees. Cut regularly to control the size for safety reasons and to reduce the risk of soil shrinkage that might cause subsidence of buildings.

Lapsed pollards

There are considerable numbers of pollards, and coppice, that have not been cut for many years. As the importance of fuel wood and fodder declined and coal became more widely available the need for actively managed wood-pasture decreased. This resulted in lapsed pollards, ie those that have not been cut for many years. The branches have grown for many more years than would have been the case in the past and have become large and heavy (Figure 6). The trunk or bolling of the tree may not be able to support the weight and it becomes vulnerable to wind damage either lifting the root plate or splitting the bolling. This presents one of the most difficult management problems, which will be addressed in chapter 4.

Figure 6. *See colour plate page 83.*

Wood-pasture

Silva pastilis *or wood-pasture (Figure 7) is distinguished from* silva minuta *(underwood) in the Domesday Book. In most wood-pastures the trees were actively managed.*

A wood-pasture can be defined as a land use combining trees and grazing animals (either stock or deer) where often (not always):

- The trees are old and at low density.
- The trees are frequently managed by pollarding.
- The grazing tends to be long and sustained, leading to a different structure and species composition than ungrazed woods in similar soils.

Wood-pastures vary between very open and very dense, and three broad types are found:

- Grazed high forest with woodland type flora.
- 'Parkland' with a ground flora showing few woodland elements.
- Grazed coppice in which livestock are temporarily excluded until the regrowth is out of reach.

Wood-pastures that are no longer grazed are termed 'former wood-pastures'.

Figure 7. *See colour plate page 83.*

Why ancient wood-pastures are good for wildlife

- They tend to have a wide range of tree age classes with veterans well represented (even though the veterans are often the result of management).
- They tend to have a mosaic of glades, open and dense woodland.
- The tree boles are often well lit and not heavily shaded by scrub or brambles (unlike ungrazed woodland), a condition favoured by many species.
- There tends to be a high quantity of dead and dying wood on the living trees.

These conditions are better represented in wood-pasture than other modern managed woods. Sites with a combination of wood-pasture and old growth woodland tend to be the most valuable in terms of nature conservation.

2.3.3.2 Regional variations

Pollard form varies between regions owing to different management. In more northerly countries the importance of the trees for winter fodder (from the leaves and bark) was greater and in many situations the land under the trees was used for making hay (see Bergendorff & Emanuelsson 1996 and Hæggström 1998). In northern areas of Britain this practice of pollarding was sometimes called cropping. Leaves from pollards may have been used as fodder less frequently when agricultural techniques provided a wider range of winter fodder crops. In recent times hedgerow trees were pollarded in Nottinghamshire as a 'last resort' for cattle in bad years (N. Lewis pers. comm.). Holly is also currently cut in the New Forest for winter pony fodder and in Killarney for sheep.

2.3.4 Other types of veteran trees

In addition to the three main categories several other types may be found (see also Figure 8).

2.3.4.1 Bundles

The term bundle is used to describe a tree, which by design or accident has originated from two or more seedlings or plants grown in close proximity. Bundles are normally, but not always, of the same species. As the young trees grow the individuals become very closely pressed together. Some single boles show natural fluting and convolutions and it is rarely possible in single species groupings to be confident of their origin by visual inspection. Because of the way that they grow, bundles often have many of the characteristics associated with veteran trees. Reasons for planting bundles are not always known but broadly speaking three main types can probably be distinguished:

- a naturally occurring bundle, the result of an accident of seed fall or an animal burying a cache of seeds that then germinate;
- a forester planting trees who slips several in a hole together to finish the task quicker;
- the result of a planned decision to create a bundle or multi-stemmed tree. This can be for several reasons, for example:
    - for landscape purposes, often in designed landscapes to create a wide spreading crown more quickly. For example, it was recommended by Evelyn in the 17th century and is a technique known to landscapers;.
    - for agricultural purposes. In some wood-pastures a few bundles can be found. This may of course be accidental but it has been suggested that they might have been deliberately managed to confer distinct benefits, eg produce seed (when all the other trees around them were pollarded regularly and did not).

2.3.4.2 Fused coppice stools.
These are abandoned coppice stools where the stems have grown close enough together to have fused for some distance above the original stool. They can be difficult to distinguish from bundles.

2.3.4.3 Shredded trees.
A tree where the side branches are cut back repeatedly with a small tuft sometimes retained at the top of the tree. These are now very rare in Britain, though relics occur in the New Forest (N. Sanderson pers. comm.), but they still occur in other countries such as France. Most shreds are probably not particularly old.

2.3.4.4 Coppards.
Trees coppiced and then later pollarded (or bundle planted trees later pollarded), a feature of parts of Epping Forest (Essex) and Dalkeith Old Park (Mid Lothian).

2.3.4.5 Singled coppice stools.
A coppice stool where one limb has been retained, when the others were cut, and is left to grow on as a tree (ie is stored).

2.3.4.6 Layered trees.
Layering is a means by which some tree species naturally regenerate. Old trees may fall over completely and then re-grow or collapse and layer well away from the original base. This is characteristic of lime, willow, alder, black poplar, medlar and bird cherry but can occur in any species. A particularly notable example is the Tortworth Chestnut. The term phoenix regeneration has been applied to trees that have fallen over, or split apart, and successfully continued growing.

2.3.4.7 Orchard trees.
These trees are pruned to encourage fruit production and for ease of management but the act of pruning will enable the trees to live longer than they otherwise might. Veteran orchard trees have a very distinct invertebrate fauna associated with them.

2.3.4.8 Naturally damaged trees.
The effect of browsing, wind, fire, grey squirrel or oak scale insect damage can act in a similar way to pollarding. Usually these events shorten the life of the tree but they can create similar conditions to those found in veteran trees. Where the top of a tree has been removed by an agent other than man it is often referred to as a natural pollard or having been self-pollarded.

The categories are differentiated according to management practices and if these have changed or ceased for long periods it becomes difficult, if not impossible, to assign trees confidently to a particular category. For example, a tree pollarded once or twice, then grown on for over 50 years may be indistinguishable from an open grown maiden with a multi-stemmed crown.

Figure 8. *Diagram to show some of the types of veteran trees.*

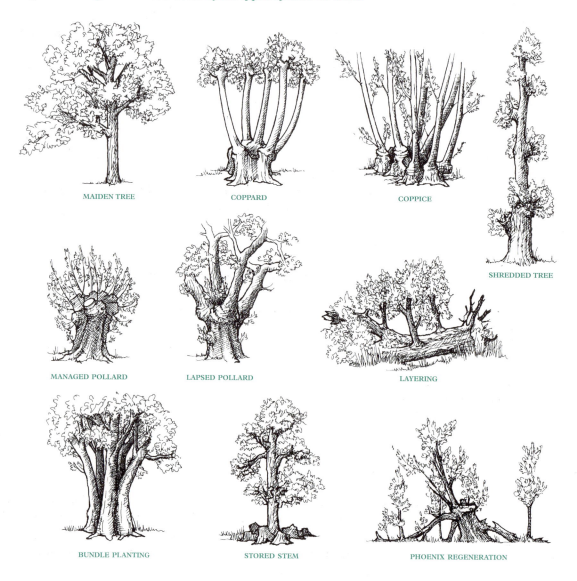

MAIDEN TREE COPPARD COPPICE

SHREDDED TREE

MANAGED POLLARD LAPSED POLLARD LAYERING

BUNDLE PLANTING STORED STEM PHOENIX REGENERATION

2.4 Threats to veteran trees

There is a tendency to view old trees as immutable and immortal. They have demonstrated their resilience to past threats but some of the potential threats of today have no precedents or are on a scale, or are taking place at a rate, that may outstrip the ability of the trees to adapt. Vigilance is needed to identify future threats. Those most frequently encountered today are:

- felling - to obtain the wood and timber, for safety reasons, to increase tidiness, for change in land use (eg development or agriculture) or for landscape reasons;
- competition from surrounding trees both planted and naturally occurring (or sudden release from competing trees);

- neglect (lapsed pollards having heavy branches that the tree is unable to support);
- inappropriate past management (eg filling cavities with concrete, girdling with chains and iron bands);
- unskilled tree surgery (eg cutting into the bolling, uncontrolled major limb removal, damaging retained limbs);
- inappropriate management of surrounding land (eg ploughing close to the trees, use of agricultural sprays and fertilisers or damage to roots by development, trenching and cable installation);
- inappropriate grazing levels (too little results in tree cover that can shade out the old trees, too much does not allow any tree regeneration and can lead to bark stripping, soil compaction, enrichment etc.);
- rapid changes in water table levels or surface water causing drought (eg owing to increased abstraction or naturally induced) or water-logging owing to raised levels;
- fire - externally, eg through fires in the surrounding land, bonfires, or internally owing to vandalism;
- pollution - remote, from industry and traffic, or localised, eg toxic rubbish such as oil and chemicals close to the tree, salt on roadside trees or nitrogen enrichment owing to manure and compost heaps;
- trampling/soil compaction - caused by livestock, people or vehicles (see Figure 9);
- bark damage - caused by people, vehicles or livestock;
- disease - eg Dutch elm, oak dieback;
- lightning strike.

Often some of these threats are accelerated when land changes ownership.

Populations of old trees and their associated wildlife are also threatened, in additional ways by:

- isolation and fragmentation;
- lack of a new generation of old trees;
- removal of standing dead trees and dead wood.

Figure 9. *See colour plate page 83.*

2.5 Why manage veteran trees?

Management of veteran trees is often needed to ensure that the threats, identified before, do not cause loss of the trees and the value associated with them. Active management may not involve doing very much for most of the time. **The essential point is that the trees and their situation are checked at regular intervals and management carried out only if it is necessary.** Each situation must be assessed individually.

The broad reasons for managing old trees have changed quite considerably over the years. From Neolithic times until the 18th century the chief reason for the management of woodlands and trees was for their wood, timber, bark, leaves etc. In the 18th and 19th centuries the recreation of the wealthy started to become an influence and people desired places for quiet walking, picnics and exercise. In the 20th century reasons for management have changed again. As the need for small-scale wood (pollard products) declined many existing trees were just neglected. Others were removed in the process of urbanisation and agricultural intensification. Those that remain have become valued for amenity and biological values. Coupled with this, an interest in repairing landscapes has developed, especially designed ones.

In the last decade the importance of biological value has been further focussed by the Rio Convention on Biological Diversity and so current management aims to provide a continuation of habitat. If no management is carried out habitats associated with veteran trees will be lost. Dependent specialised species of limited mobility will die out. Management may also take place for landscape, economic, or cultural reasons or a combination of several of these. An opposing pressure has come from a different quarter; as Britain becomes an increasingly litigious society the 'management' of trees for safety reasons has also grown substantially.

Reasons for managing old trees:

- to safeguard the genetic resource;
- to provide continuity of habitat for wildlife;
- to keep individual trees alive for as long as possible, enabling a new generation of trees to replace the old ones;
- to maintain traditional practices;
- to perpetuate maturity and continuity within landscapes;
- to perpetuate aesthetic values eg characteristic landscape features;
- to increase the landscape value;
- for historical reasons - association or landmark trees;
- to fulfil safety responsibilities.

Continued management today and into the future depends upon those who have stewardship of veteran trees acknowledging their present value and ideally finding new values.

Further reading: Alexander, Green & Key. (1996), Barwick (1996), Bergendorff & Emanuelsson (1996), Damant (1996), Debois Landscape Survey Group (1997), English Nature (1996), Green (1994, 1995c, 1996a, 1996b), Hæggström (1992, 1994, 1998), Le Sueur (1931), Peterken (1996), Pott (1989), Quelch (1997), Rackham (1986, 1991), Rush (1999), Sanderson (1998a, 1998b), Smout & Watson (1997), Watson (1997).

| Chapter 3 | How a tree grows, becomes old and decays |

3.1 Tree growth

This section outlines the principles of tree growth that have consequences for management techniques. It is not the intention to give a detailed account of the physiology of trees.

3.1.1 Tree growth

A cross section of a tree (Figure 10) illustrates some of the features important for growth. The **bark** forms a protective, waterproof layer, and actually consists of several layers, the innermost of which is the **phloem**, which transports food from the leaves to the rest of the tree. Inside the phloem is the **cambium**, which is the region of growth, or meristematic, cells. These cells divide, forming phloem to the outside and **xylem** to the inside. The xylem is where the water is transported from the roots to the leaves and forms the wood of future years. The outer bark usually remains a relatively thin structure (although it can compose up to 10% of the radius in veteran trees) but the wood builds up so that the overall girth of the trees gets bigger each year as well as the tree, usually, increasing in height. Trees are not perfect cylinders however; they taper towards the top and the higher up a tree a cut is made across it, the fewer the rings that can be counted.

Recently formed xylem (sapwood) consists of conductive pipes surrounded by living parenchyma (packing) cells. In some tree species (eg beech) the living cells progressively die over a period of years and the tissue becomes non-conductive. This older, non-conductive wood is then called **ripewood**. In other species (eg oak) the living xylem cells are genetically programmed to die after a certain period of time (approximately 10 years in oak) and after this is termed **heartwood**. Heartwood may contain substances that increase its resistance to decay. The variation in wood formation and structure between different tree species has consequences for the rotting processes and the organisms associated with rotting and also the longevity of the tree.

3.1.2 Annual growth

Living trees always add annual increments of sapwood, although their width may vary according to growing conditions and the age of the tree. Trees in Britain hardly grow in the winter months. In the spring they grow very fast; the wood produced has large cells with thin walls and is the **earlywood**. Later in the year, when the growth is slower, **latewood** results, with smaller cells and thicker cell walls. These differences in growth are seen as rings in the wood when a tree is felled. (False rings can occur some years due to **lammas** (late summer) **growth** or after a stressful weather event such as a drought). The relative widths of the annual rings can give an indication of the growth rate of the tree in a particular year. There may, though, be variation between branches on the same tree eg one side may be growing more quickly than the other, and this can give rise to eccentric rings.

Pollarding has a considerable impact on annual rings and tree growth. After cutting, the crown is reduced in size, so for the first few years the trunk of the tree expands slowly and the rings are narrow. The width of the rings gets gradually wider (varying of course with other local conditions) until either the tree is cut again, or it resumes the growth rate of a maiden uncut tree.

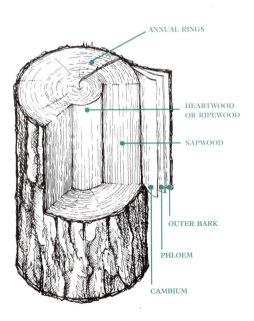

ANNUAL RINGS

HEARTWOOD OR RIPEWOOD

SAPWOOD

OUTER BARK

PHLOEM

CAMBIUM

Figure 10. *Diagram of the internal structure of a tree.*

3.1.3 Forming new branches

There are two different processes by which a tree can produce new branches from old stems: from dormant buds (also called epicormic growth) and as adventitious growth. In addition the growth of existing small stems may increase, relative to others, following tree surgery or damage.

3.1.3.1 Epicormic growth from dormant buds (Figure 11)

Dormant buds form from the growing stem or branch of the tree, but do not develop any further at that time. They become embedded in the bark but, by growing a small amount each year they are able to maintain their position and not become completely engulfed by the wood as the tree expands in thickness. Sometimes they can divide to form additional buds that also remain dormant.

Hormones, such as auxins, from the crown of the tree probably keep the buds in a suppressed condition, but if some change in the root to foliage ratio occurs in the tree this alters the balance of the hormones and the dormant buds may start to grow. The types of change that stimulate growth include ring barking, severe pruning, exposure of the tree to increased light levels and water-logging.

Some species of tree lay down more dormant buds (eg oak, lime, English elm, poplar species, ash) than others (eg beech). Dormant buds can survive in a suppressed condition for many years and then grow when conditions are favourable. However their viability does decline over long periods of time. The longevity of buds is believed to be in the region of 100 years for oak, 60 years for hornbeam and sweet chestnut and less than this for beech and willow. This is one of the reasons why old trees are less able to respond to cutting than young ones. It is also thought that, as the trunk of the tree has only a fixed number of dormant buds, repeated pollarding will eventually exhaust the supply. Trees cut repeatedly at short intervals, such as street limes and planes have shown a decline in response to cutting over a long period of time, which may be due to the fixed number of dormant buds. Thus, leaving some young growth on the tree may increase the chance of viable dormant buds being present and hence the chance of regrowth. Dormant buds are not usually distributed evenly over the surface of the tree but form in clusters. Rough bark or burrs may indicate a higher density and trees that have these features may respond better to cutting than smooth barked trees. Some epicormic shoots grow from the buds as soon as they are formed, ie the buds do not have a dormant period.

In many situations epicormic growth is viewed as a detrimental characteristic. Where shoots occur they cause knots in the wood and this reduces the timber value of trees such as oak. They are potentially hazardous in street trees where lots of small branches projecting from the main stem can damage cars and hurt pedestrians. There may also be a genetic component in the production of such growth habits. **For the success of pollarding, however, the more dormant buds the better the chance of regrowth and survival.** Perpetuating the genetic stock of trees on sites where pollarding has been carried out in the past may be better than planting commercial stock, which is likely to have been selected because it produces high quality timber with few knots and therefore few dormant buds.

3.1.3.2 Adventitious growth

Adventitious buds form when a tree is damaged. They result from injury or pruning, developing from the callus tissue that forms at the point of damage. Typically a cluster of small shoots develops but it is unusual for them to persist for many years.

While growth from dormant buds arises from a deep-seated connection to the trunk of the tree, adventitious growth is much more superficial. As a result it is not as strong and is more easily broken. The presence of adventitious growth on old trees is encouraging but often of less value than growth from dormant buds in the long term. Again, some species of tree are better at producing adventitious growth than others. Oak is generally poor, beech is often cited as being good, but recent experiences at Burnham Beeches and Epping Forest with both old and young trees has not borne this out. Adventitious growth *may* develop better from natural tears rather than saw cuts, owing to the increased exposure of the cambium.

3.1.3.3 Growth of existing branches

Crown reduction in old trees may produce a third form of growth as a result of light reaching retained branches that were previously receiving low levels of light. As a consequence, small existing shoots grow rapidly into the light. As the years progress, they become the major branches. This type of growth is characteristic of trees that generally respond poorly to being pollarded (eg conifers and beech), and this is how the classic candelabra-shaped beech pollards arise.

Figure 11. *Diagram to show how a branch grows.*

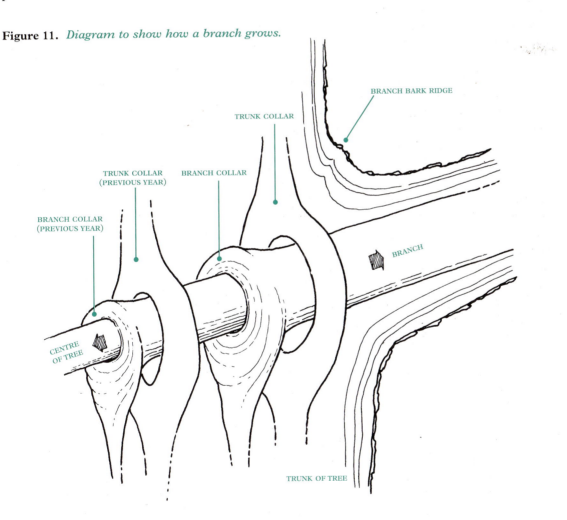

3.1.3.4 Repeated cutting

Repeated cutting back to the same point may result in swollen areas. This can be seen in old trees and also younger street trees which have been cut many times. This has been attributed to the active growth points attracting a good supply of food, which results in excessive wood production and/or reaction wood, the result of loading from developing branches.

3.2 The stages in the life of a tree

Trees do not have a fixed life span; some die before reaching veteran status, others will become veterans at a much earlier age than might be expected. There is considerable variation both between and within different tree species. The life of a tree in natural conditions may pass through three main stages (Figure 12):

1. **Formative** - *This is the stage when most of the energy produced by a tree is used for growth. There is a rapid increase in size as it grows from a seedling to a fully mature tree. Crown size and leaf area increases each year, until the canopy is fully developed. The widths of the annual rings are similar each year but because the whole tree is getting bigger, the cross-sectional area covered by each successive ring is greater.*

2. **Full to late maturity** - *This starts when the optimum crown size is reached. The amount of food produced from the leaves remains much the same each year and results in a more or less constant volume of wood being laid down. However, as the tree gets ever larger, this volume is spread increasingly thinly, thus the rings in the stem decline in width.*

3. **Ancient (Veteran stage)** - *This is the stage reached when the successive increments added to the tree, seen as the rings of wood, have a reducing cross-sectional area, but the tree is still increasing in girth. The crown dies back and branches may be lost, damage and decay also reduces productivity. The result is that as the leaf area declines, less new photosynthetic material is produced each*

Figure 12. *The stages in the life of a tree.*

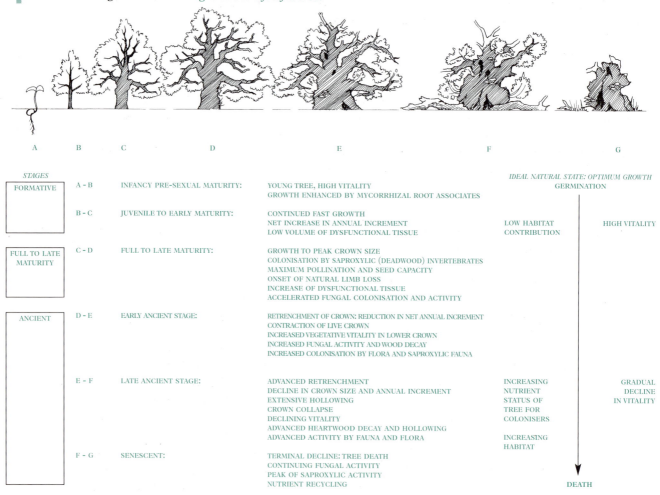

STAGES					
FORMATIVE	A – B	INFANCY PRE-SEXUAL MATURITY:	YOUNG TREE, HIGH VITALITY GROWTH ENHANCED BY MYCORRHIZAL ROOT ASSOCIATES		*IDEAL NATURAL STATE: OPTIMUM GROWTH* GERMINATION
	B – C	JUVENILE TO EARLY MATURITY:	CONTINUED FAST GROWTH NET INCREASE IN ANNUAL INCREMENT LOW VOLUME OF DYSFUNCTIONAL TISSUE	LOW HABITAT CONTRIBUTION	HIGH VITALITY
FULL TO LATE MATURITY	C – D	FULL TO LATE MATURITY:	GROWTH TO PEAK CROWN SIZE COLONISATION BY SAPROXYLIC (DEADWOOD) INVERTEBRATES MAXIMUM POLLINATION AND SEED CAPACITY ONSET OF NATURAL LIMB LOSS INCREASE OF DYSFUNCTIONAL TISSUE ACCELERATED FUNGAL COLONISATION AND ACTIVITY		
ANCIENT	D – E	EARLY ANCIENT STAGE:	RETRENCHMENT OF CROWN: REDUCTION IN NET ANNUAL INCREMENT CONTRACTION OF LIVE CROWN INCREASED VEGETATIVE VITALITY IN LOWER CROWN INCREASED FUNGAL ACTIVITY AND WOOD DECAY INCREASED COLONISATION BY FLORA AND SAPROXYLIC FAUNA		
	E – F	LATE ANCIENT STAGE:	ADVANCED RETRENCHMENT DECLINE IN CROWN SIZE AND ANNUAL INCREMENT EXTENSIVE HOLLOWING CROWN COLLAPSE DECLINING VITALITY ADVANCED HEARTWOOD DECAY AND HOLLOWING ADVANCED ACTIVITY BY FAUNA AND FLORA	INCREASING NUTRIENT STATUS OF TREE FOR COLONISERS INCREASING HABITAT	GRADUAL DECLINE IN VITALITY
	F – G	SENESCENT:	TERMINAL DECLINE: TREE DEATH CONTINUING FUNGAL ACTIVITY PEAK OF SAPROXYLIC ACTIVITY NUTRIENT RECYCLING		DEATH

Figure 13. *Sweet chestnut trees at Croft Castle Herefordshire). A standing dead tree is in the foreground and stag headed trees behind.* *(see colour plate page 84).*

year and the tree is even less able to maintain a complete cover of woody material over the whole stem area. This process is called retrenchment and is seen most visibly as 'stag-headed' trees (Figure 13), typically in oak. This does not mean that the tree is about to die, it is a condition that can persist for many decades or even centuries. Retrenchment is not the only cause of stag-headed trees, it can also occur in younger trees, brought on by drought, disease, insect damage, root disturbance or pollution. The response of the tree results in a new balance between the area of woody material and that of the leaves. A tree in the last phase of its life that has retrenched can be very healthy and vigorous despite extensive decay and dieback. This stage may be also be the longest in the life of the tree.

The ancient stage can be further subdivided into three phases.
- **Early ancient.** When, over a period of years, there is a trend for the amount of dieback to exceed growth.
- **Mid-ancient.** When the annual rings cannot form all the way round the stem and some discontinuities start.
- **Senescent.** The terminal decline of the tree, leading to death.

Tree species vary in the proportion of time they spend in each of these phases. Willow and birch tend to have an extremely short 'mature phase' whereas others, such as yew can grow in cycles, passing from ancient back to formative growth. All the stages are a continuous process and of variable length. Once a tree has reached the middle ancient stage, nothing should be done to encourage the speeding up of the ageing process and the aim should be to keep it in this phase for as long as possible. When discontinuities in the annual rings develop, the tree is at its most vulnerable stage.

3.2.1 Other aspects of the veteran stage

Veteran trees that are retrenching tend to show a diminished growth rate and a drop in reproductive output. They are also slower to occlude wounds if damaged. They tend to develop other features and characteristics to a greater extent than younger trees (eg cavities in the trunk, seepages, dead loose bark, dead wood in the canopy and physical damage). One important point to note is that, **as the tree ages it becomes more valuable for a wide range of other organisms and its habitat value increases.**

3.3 Assessing the age of a tree

Assessing the age of a veteran tree is not an easy task and is usually, at best, an estimate. There are a number of methods that can be used, however.

- **Taking a core** is one option but many veteran trees are hollow or rotten to a greater or lesser extent. Coring is also detrimental as it can cause damage in an undesirable place (Shigo, 1986a illustrates the possible consequences of taking a core from a tree in terms of the rot induced.) Felling a veteran tree to count the rings should, of course, never be done! Counting the rings on major branches (cut during the course of work or fallen) from veteran trees can sometimes give an indication of age. An allowance of years for the tree to produce the branch should be added to the ring count.
- **Age based on tree girth I.** Because of the variation in growth rate throughout the different stages in the life of a tree, caused by differing conditions of soil type, nutrient status, rainfall etc and the even greater variation between different species it is difficult to extrapolate age from girth measurements. There are some rules of thumb that can help. Mitchell (1974) states that one inch (25 mm) of girth (at breast height, 1.2 m from the ground) is equivalent to one year's growth for a free standing tree and 1/2 inch (13 mm) in a tree within a woodland setting. This approximation of age is only helpful for some species of tree (eg oak) that are in middle age and have not been cut. It is of very limited value with regard to old trees.

- **Age based on tree girth II** This is a more accurate (and thus more complicated) system also based on tree girth developed by J. White. It has been calibrated using a variety of older trees where the planting date is known and provides the best estimates of age available. (For a description of methods and the relevant tables used for calculation see White, 1998.)
- **Age/size ratios within species.** Different species of tree grow at very different paces but it is possible to build up a table of girth sizes and estimated age for a particular species, though even within a species there is considerable variation. Yew has been studied with this aspect in mind (Baxter 1992).
- **Site-specific information.** It is possible to draw up a table for a particular species on a particular site that allows figures for girth to give an estimate of age and which may be able to take into account the effects of historical management such as pollarding (eg Le Sueur 1931). These may not correlate well with data from other sites.

As a general principle it is almost impossible to age pollards, or trees that have undergone natural crown loss, by the methods given above.

3.4 Why does pollarding or cutting cause trees to live longer?

A normal tree reaches the veteran stage when the size of the crown is not large enough to produce enough food to maintain the same cross-sectional area for each annual ring. During the process of retrenchment the photosynthetic area is reduced, as is the surface area of the woody branches, so that less food is required by the tree.

Reducing the size of the crown at intervals delays the veteran stage in a tree's life when the demand for water and nutrients outstrips its ability to increase the root area to absorb them. A reduced crown also reduces the risk of wind throw, owing to the relatively low stature (short 'lever-arm') and small 'sail area', but a long abandoned pollard with a large sail area may become particularly vulnerable.

In addition, the multiple branches from the top of the bolling produce a larger number of vascular connections into the trunk than in a normal tree. These, in effect, form separate compartments and it is less easy for pathogenic agents, or aggressive decay fungi, to spread through the entire tree.

Severe wounding of a tree is similar to pollarding and has four main effects:
- exposure of cut surfaces to micro-organisms;
- drying out of wood from the cut surfaces;
- reduction in the volume of foliage and sapwood in the tree and thus its existing stores of carbohydrates and the capacity to replace them;
- loss of shoot tips which disrupts hormonal co-ordinating signals and affects growth.

Wounds result in the drying out of an area of wood, causing decay. The larger the amount of wounding (as on an old tree with all branches removed) the more drying out and dieback with the increased chance of infection by micro-organisms. The ability of the tree to compartmentalise (see section 3.5.1) will be reduced because of its severely reduced photosynthetic area. It seems that a few trees are able to cope with this situation although willows can grow new shoots from an old bole. If some branches are retained on the tree, the amount of exposure, drying out, and infection by micro-organisms is decreased. However, these areas will be restricted to strips of xylem and phloem associated with the cut branches. For this reason (and to keep the sap wood active) it is best to retain good connections of xylem and phloem throughout the tree, thus maintaining 'channels' of living tissue between the roots and shoots. In some old trees that have been cut back heavily on one side this connection has been broken and the tree has died back completely thus resulting in a 'lop-sided' tree. Leaving branches, at least small ones, all round the bolling is therefore advisable.

VETERAN TREES
I N I T I A T I V E

> ### *The value of pollarding and working trees*
>
> *When humans started clearing the forest they removed many of the conditions that saproxylic species (or wood decay communities) required such as dead wood on the ground and within trees and standing dead trees. While this must have been detrimental to many populations of saproxylic species the human management of the trees eg pollarding and coppicing, created very similar conditions in a different way. Almost all trees may, at one time, have been used or managed and these can be referred to as 'working trees'. The increased life expectancy of working trees and the characteristics they developed enabled the perpetuation of suitable niches for a range of species in the wood decay community despite the change in the landscape that resulted.*

3.5 The decay process

The process of decay in wood is a complex subject and the details are only just starting to be understood. There are many different agents involved, which make it very difficult to establish the relative importance of each. What is clear however, is that fungi have a fundamental role in the process. The work of A. Rayner, L. Boddy, A. Shigo, F. Schwarze and D. Lonsdale has helped enormously in establishing how fungi behave within living trees and dead wood and how the tree responds. The following sections summarise the crucial points with respect to living trees. Section 7.5 considers the conservation of fungi in relation to veteran trees.

3.5.1 Compartmentalisation

Trees have no wound healing processes, as animals do, but they do have a way of limiting any damage caused. If a tree has been damaged and is then cut some years later it can be seen to have dried out, the dysfunctional area of wood extending back from the wound. This area often has a sharp boundary wall between it and the rest of the tree as shown by a difference in the colour of the wood (Figure 14). This process of boundary setting has been termed compartmentalisation.

The sharp boundary results from a response of living cells to the ingress of air and/or micro-organisms and may represent a barrier between healthy and damaged areas. If a tree is badly damaged it spends energy in compartmentalising, leaving less for growth, which can result in a smaller annual ring. The more areas that are 'sealed off', the less tissue is available for the tree to distribute food and water to its various parts. Eventually, when there are too many dysfunctional compartments and the distribution of new sapwood becomes discontinuous, the tree is unable to maintain vital functions and death results. However, the more compartments there are in a tree, the more structural diversity there is and so the larger the number of niches and habitats for other organisms. The exact processes by which compartmentalisation and barrier formation occurs remains unclear.

Figure 14. *See colour plate page 84.*

3.5.2 Fungal colonisation

It seems likely that fungi colonise living trees in two main ways:
- **from the outside. In the simplest scenario, physical damage to a tree weakens its physical defences and makes conditions suitable for the fungus to colonise, become established and grow;**

- from the inside. The fungus makes use of the tree's own plumbing system (xylem and phloem) to reach different parts of the tree via the sap stream; this can occur at any stage in the life cycle. The fungi often remain in a latent (inactive) state without any noticeable impact on the tree until conditions within the wood change enough to activate them, eg drought, ageing process.

It is thought that the sapwood of a healthy tree has such a high moisture content that it is unsuitable for the growth of most fungi. However, when the tree is mechanically damaged or is stressed in some way, parts of it may become more suitable for fungal growth. Loss of a branch, for example, allows air in and causes drying out of the wood around the wound and enables fungal growth. Stress brought on by drought or the severing of roots may cause the tree to stop producing food (photosynthesising) from a branch. This branch then dies back and dries out because the flow of sap is no longer as strong as normal. The drier conditions activate some of the latent fungi or fungi entering via the dead or broken wood.

Most of the fungi capable of causing extensive decay depend on wounds or dead branches or roots as entry points. Some of these species grow only in heartwood, while others are confined to sapwood or are able to colonise either. A wide range of factors determine whether or not decay becomes extensive enough to weaken the tree significantly.

Some pathogenic fungal species are able to cause death or dysfunction to parts of the tree even without stress or major injury (eg some honey fungus species or the fungus that causes Dutch elm disease). This relative minority of species are a primary cause of dysfunction in the sapwood or of death of the cambium.

As the fruiting bodies of the fungi are the only parts that are usually noticed they are often misinterpreted. A small number of species (eg some species of honey fungus) can cause the death of a tree but a much larger number produce fruiting bodies only when the tree (or that part of it with fungal fruiting bodies) has died from other causes (ie they are saprophytic). This leads to many misconceptions as to the role of fungi.

The means by which fungi colonise sapwood

- **Root colonisers.** Species that colonise intact roots and then spread throughout the cambial zone of the tree. They may kill the tree by girdling it or killing too many roots, eg *Armillaria mellea.*
- **Sapwood colonisers I.** Species that enter the tree through a wound or other open entry point on the tree. Most species decay parts of the tree without killing it but sometimes the decay parts of the tree without killing it but sometimes the decay is so extensive that very little functional sapwood is left, eg most *Ganoderma* species.
- **Sapwood colonisers II.** These species also enter the tree through wounds but are more aggressive and may kill the host, eg *Chondrostereum purpureum.*
- **Deadwood colonisers.** Species that can colonise sapwood only after it has died as they are unable to overcome the active defences of the tree, eg *Daedaleopsis confragosa.*

Note that the behaviour of fungi covers a spectrum and that a particular species may fall in between the categories presented here.

3.5.3 Fungi growing within the heartwood

Some fungal species are able to grow in the innermost part of the tree, which consists of dysfunctional wood. It is usually drier than the outer sapwood and so is more suitable for the growth of fungi if they are present. Species that rot the heartwood such as

Laetiporus sulphureus (Figure 15) break down only the dead wood. This decays the centre of the tree but leaves the outer, living layers intact. While this may not be desirable from the point of view of a commercial forester, the tree is not harmed and may actually benefit. Decay and hollowing are part of a nutrient recycling process. The tree can make use of the products of wood decay within the trunk by producing aerial roots from its above ground parts, which grow into the rotting stem. A hollow tube may respond differently from a solid trunk in high winds and is not necessarily more likely to snap provided its walls are not so thin that buckling occurs.

Figure 15. *See colour plate page 84.*

3.5.4 Types of decay

There is a wide range of variables influencing decay. The result of this is a tremendous range of potential niches available to organisms such as invertebrates that make use of the rotting process and its products. Premature decay in a tree is not necessarily detrimental, either to the tree or to its wildlife value. Young decaying trees can be very valuable, on sites with veterans, in providing suitable conditions for the saproxylic organisms. Decay is dependent on many factors:

- age of tree;
- presence of heartwood. (Those species of tree, eg birch and beech lacking durable heartwood tend to decay quicker that those that do, eg oak);
- type of wound or stress agent;
- species of fungi involved and stage of growth within the tree;
- species of invertebrate involved;
- species of vertebrate involved;
- position of wound;
- whether the wound collects water or not;
- whether the wound is enclosed or open to the air;
- whether the wound is permanently covered by water (becoming anaerobic);
- ability of the tree to respond to damage (ie to form reaction zones and to occlude wounds);
- outside factors (eg dung, rotting carcasses, aerial pollution).

There are three main types of rot caused by fungi:

- **White rot** - When the lignin and cellulose are both broken down. In **simultaneous white rot** the lignin and cellulose are broken down at approximately the same rate causing loss of both stiffness and strength, which, in the advanced stages of decay produces a thick porridge-like substance. In **selective delignification** (or **stringy white rot**) the lignin is broken down first and the cellulose degrades more slowly. Initially the result is soft material that is still quite strong, the colour and weight of balsa wood. White rot is more common is broadleaved than coniferous trees, eg rot produced by some *Ganoderma* species.
- **Brown rot** - When the cellulose is degraded and the lignin is left intact. The initial results of the decay are brittle but rigid. It does not bend much before breaking but may break into cubes known as cubical brown rot. Eventually a rich, humus like, substance may result (red wood mould) usually after having passed through the guts of many invertebrates. Brown rot is more common in conifers than broadleaved trees. It is produced by, for example *Fistulina hepatica* and *Laetiporus sulphureus* in oak trees.
- **Soft rot** - This is when the cellulose is degraded, as in brown rot, but the fungi invade the cell walls in a very different way. Many white rots and some brown rot fungi can behave like soft rot fungi in living trees, but 'classic' soft rots are caused mainly by specialised ascomycetes which grow in the surface layers of dead wood or timber under very wet conditions.

Different invertebrate species and communities are associated with each of these types of rot.

3.5.5 The value of dead wood

It is important to encourage a variety of types of rot so that suitable conditions are provided for a range of the more specialised invertebrates. The more dead wood a tree contains the more valuable it is. Thus, a living veteran tree is better than a dead one because it will continue to produce more dead wood. Old dead trees left standing are usually better than those in younger growth phases. Damaged young trees may also have valuable areas of rot (natural or even artificially induced).

Trees containing a higher volume of wood have a higher wildlife value, which is why old pollards are generally more valuable than old coppice stools. The latter may have a range of niches, but the sheer volume of wood is considerably less than in most old pollards.

3.5.6 The role of organisms other than fungi

Although fungi have the fundamental role in the decay process in trees, they are not the only active organisms. Many invertebrates assist in the breakdown of wood by boring into it and feeding on the comparatively softer and more nutritious bits. They also enable fungal mycelia to penetrate the wood more easily along the sides of burrows. Some species of insect have nitrogen-fixing bacteria in their guts, which enhance the nutritional value of their faeces which may be re-ingested by other species.

Wood has a very complex chemical structure and is very indigestible. Many of the invertebrates rely on fungi to break the wood down into simpler molecules so that they can take advantage of it. Ambrosia beetles (family Scolytidae) even have fungi associated with them, which they carry between trees to perform this function.

Birds such as woodpeckers may contribute to the process by actively hollowing out areas for nesting, their nests and holes may then be inhabited by other animals. The faeces and dead bodies that build up in the tree holes contribute to the nutrient status of the rotting wood. There are even secondary fungal colonists whose fruiting bodies are found in cavities created by the primary decay fungi.

Further reading: Beckett (1975), Boddy & Rayner, (1983), Coder (1996), Dolwin *et. al* (1998), Graham (undated), Green (1993, 1994, 1996a), Le Sueur (1931, 1934), Lonsdale (1996, 1999), Mattheck & Breloer (1994), Mitchell, A. (1974), Mitchell, P. (1989), Patch, (1991), Patch Coutts & Evans. (1986), Rackham (1986, 1990, 1991), Shigo (1986a, 1986b), White (1996, 1998), Wignall, Browning & Mackenzies (1987).

INITIATIVE

Chapter 4   Management of Veteran Trees

4.1 Assessing the situation and planning

There are two basic things to consider when managing veteran trees. One is the individual tree and the other the site. In this chapter the tree is considered, chapter 5 covers managing the site or land surrounding the veteran tree.

This section looks first at how to assess your tree and make a decision on what to do (4.1 and 4.2). More detailed practical issues pertinent to individual trees (4.3) and the sites or landscape areas with veteran trees (chapter 5) are then discussed.

Ancient trees are found in an extremely wide range of situations. Therefore it is not possible to present a simple, easy to follow, set of guidelines which will work for every tree in every situation. One of the most frequent phrases used when contemplating management is that **every tree is an individual**. Even on a single site, different management options and prescriptions are needed for different trees. There may be several different ideal management options depending on the point of view taken by an adviser. A nature conservation officer might prefer a set of options that may be different from that of a historic landscape adviser (chapter 6) and these may contrast again with the view of a health and safety officer (see *Veteran Trees Initiative health and safety* leaflet). In addition someone with an interest in lichens (section 7.3) may recommend a different approach to an invertebrate specialist (section 7.6). Whichever option is chosen, if the tree itself is not saved, all the values associated with it are lost. Therefore it is important to know your site and your trees and to have as much information as possible on the historical background and current conservation values and status of the site which can be used to inform the management process. Potential conflicts of interest often do not turn out to be the tremendous problems that they might at first seem, when key issues are carefully considered.

4.1.1 The importance of management plans

A plan helps the manager to manage the site, and explains to others what is being done, and why. It is difficult to manage a site consistently without one. The management planning cycle can be summarised as follows:

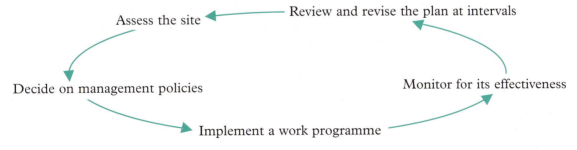

The approach you take depends on the purpose of your plan, the nature of the site and the needs of the site owner. There is no single 'right' way of doing things but it is important that all aspects are considered to make an informed judgement.

Reasons for producing a management plan:

- to ensure continuity of management over time;
- to bring together people involved in the management of the site and achieve consensus;
- to manage multiple uses and potentially conflicting interests on the site;
- to relate the site to the wider ecological and social context;

Veteran Trees: A guide to good management ■ Page 35

- to attract resources (as part of a bid for grant aid) or be the basis for a legal agreement;
- to ensure that management can be achieved within the resources available;
- to promote and publicise the site.

The plan should include separate sections which:

- record the existing attributes of the site (eg particular wildlife and historical features);
- give information about the site;
- identify the site's value and significance;
- explain what management is intended to achieve;
- outline the means that will be used to do this;
- say who will do what, when and what resources are required;
- provide a way of checking the effectiveness of the site management.

Producing a plan should not become an end in itself. If the proposals are impractical, or not clearly set out, the plan will not be used.

It needs to be decided early on who will produce the plan, whether you do it yourself or employ a consultant. Whatever the decision, the complexity of issues surrounding the management of veteran trees and the sites they are found on will almost certainly require you to seek specialist advice. More information and guidance on management planning is given in the Countryside Commission (1998) guide which includes information about a range of planning methods including the Conservation Management System (1996) which is also a computer programme for planning and reporting.

4.2 Managing individual trees

4.2.1 Management types and historical evidence

It is useful initially to distinguish two broad types of veteran trees; those that have been actively managed at some stage in their life and those that have not. In practice, the way we manage these two types of trees now may not be particularly different. However, historical management practices have been responsible for many of the veteran trees we have inherited today. Past management may also have some bearing on the way we treat them in the future. The majority of the veteran trees encountered that have been actively managed in the past are pollards. There are few written documents recording how and when trees were pollarded so the information is quite sparse. That which is available, from historical literature and current practice in areas where pollarding still occurs, is summarised in appendix 4. There is plenty of evidence concerning the intervals between cuts but little else of any help. Many of the veteran pollards in Britain today are no longer in a regular system of active management and need some restoration work. This is not a problem our ancestors had to deal with and there is unlikely to be historic literature to help.

Veteran trees today include coppice stools. Coppicing is covered in detail in a range of publications, eg Buckley (1992), Hampshire County Council (1991) and Fuller & Warren (1993), and the methods are comparatively well known. General coppicing techniques therefore are not dealt with any further here.

4.2.2 How to decide whether to actively manage a veteran tree

Take as a starting point the premise that the tree should have nothing done to it unless you can demonstrate a clear need. The decision to cut an ancient tree should not be taken lightly. Your decision should take account of the issues raised in the flow diagram on page 38.

To assess whether an individual tree is likely to respond positively to pruning, consider the following points (see also Figure 16):

- How has the tree responded in the past to minor tree surgery work?
- How have other trees of the same species on the same site (or close by) responded to cutting?
- How do trees on the same site respond to accidental damage?
- Is it a species with a 'good' reputation (like willow) or a 'bad' one (such as beech)?
- Does it have burrs, good epicormic growth or obvious abundant dormant buds?
- Is it a suitable shape, ie is it relatively easy to leave small branches close to the bolling after cutting?
- Has it been pruned before? If so, how long is it since the tree was last cut (the shorter time the more likely the response is to be good)?

Assessing the success of any trials in cutting will mean you have to leave the tree for several years. There are several examples of situations where trees responded well initially but died some years later, almost certainly as a direct response to the cutting.

Figure 16. *Diagram to show the characteristics of a veteran trees that is likely to respond to cutting and those of a tree that is less likely to respond.*

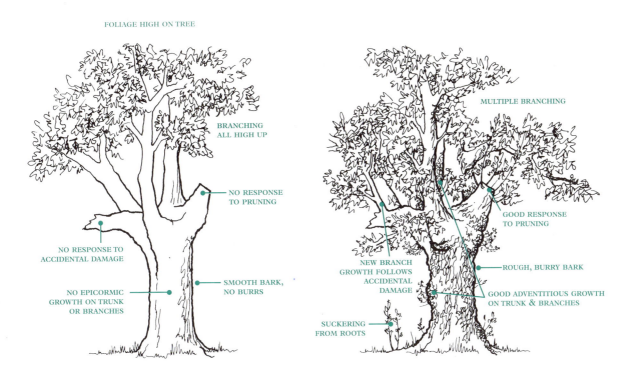

Flow diagram to help decide whether to cut a veteran tree

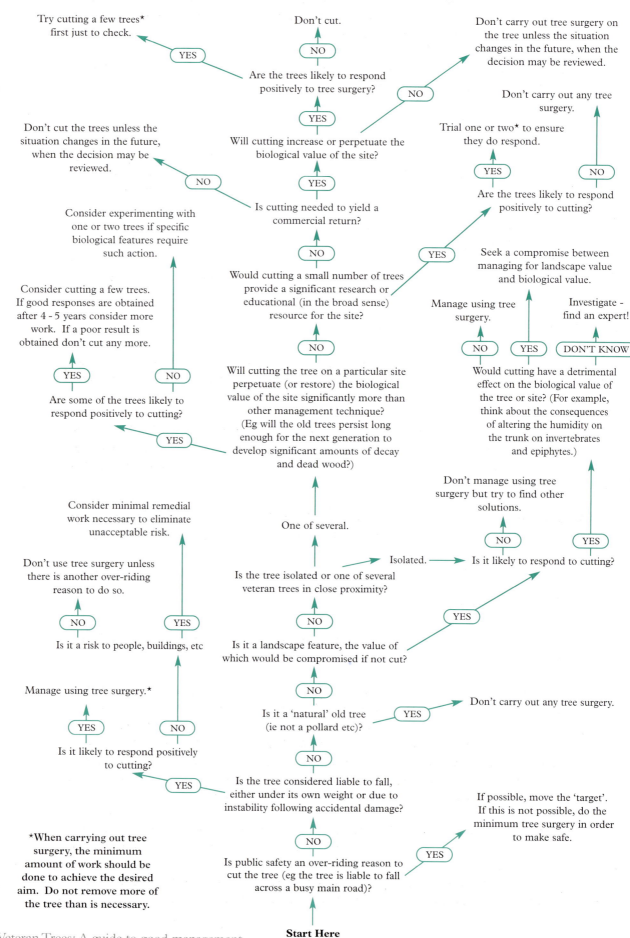

Try cutting a few trees* first just to check.

← YES

Don't cut.

NO

Are the trees likely to respond positively to tree surgery?

YES

Will cutting increase or perpetuate the biological value of the site?

YES

Is cutting needed to yield a commercial return?

NO

Don't cut the trees unless the situation changes in the future, when the decision may be reviewed.

← NO

Would cutting a small number of trees provide a significant research or educational (in the broad sense) resource for the site?

NO

Consider experimenting with one or two trees if specific biological features require such action.

Consider cutting a few trees. If good responses are obtained after 4 - 5 years consider more work. If a poor result is obtained don't cut any more.

YES   NO

Are some of the trees likely to respond positively to cutting?

YES

Will cutting the tree on a particular site perpetuate (or restore) the biological value of the site significantly more than other management technique? (Eg will the old trees persist long enough for the next generation to develop significant amounts of decay and dead wood?)

One of several.

Don't carry out tree surgery on the tree unless the situation changes in the future, when the decision may be reviewed.

NO

Don't carry out any tree surgery.

Trial one or two* to ensure they do respond.

YES   NO

Are the trees likely to respond positively to cutting?

YES

Seek a compromise between managing for landscape value and biological value.

Manage using tree surgery.

Investigate - find an expert!

NO   YES   DON'T KNOW

Would cutting have a detrimental effect on the biological value of the tree or site? (For example, think about the consequences of altering the humidity on the trunk on invertebrates and epiphytes.)

Don't manage using tree surgery but try to find other solutions.

NO   YES

Is it likely to respond to cutting?

Isolated. →

Is the tree isolated or one of several veteran trees in close proximity?

NO   YES

Consider minimal remedial work necessary to eliminate unacceptable risk.

Don't use tree surgery unless there is another over-riding reason to do so.

NO   YES

Is it a risk to people, buildings, etc

Is it a landscape feature, the value of which would be compromised if not cut?

NO

Is it a 'natural' old tree (ie not a pollard etc)?   YES →   Don't carry out any tree surgery.

NO

Manage using tree surgery.*

YES   NO

Is it likely to respond positively to cutting?

YES

Is the tree considered liable to fall, either under its own weight or due to instability following accidental damage?

NO

*When carrying out tree surgery, the minimum amount of work should be done to achieve the desired aim. Do not remove more of the tree than is necessary.

Is public safety an over-riding reason to cut the tree (eg the tree is liable to fall across a busy main road)?

YES →   If possible, move the 'target'. If this is not possible, do the minimum tree surgery in order to make safe.

Start Here

If you feel that for historical/biological/landscape reasons the tree ought to be cut in some way but that it is unlikely to respond well, it might be worth re-thinking your decision. If possible, leave it for the moment but carefully monitor what other people are doing elsewhere. In the future there may be other more successful methods of dealing with trees.

Remember, there is usually no urgency to do anything. Think long and hard before undertaking any work, and then do it only if really necessary.

4.2.3 Working on veteran trees

It is important to remember that the primary reason for working on a veteran tree is to prolong its life. Active management to increase its life expectancy may be needed because it is top heavy and about to fall over, or because a dangerous branch is overhanging a busy road (the alternative often resorted to here is to fell the tree). Generally speaking, do as little as possible in the way of cutting. There are some exceptions to this rule, for example if the intention is to return a pollard to a regular cycle of cutting again. Inducing decay and cavities in trees is worthwhile on sites where these features are scarce but experiment with these techniques on younger trees and not the veterans. Focus on keeping the veterans alive and do not do anything that might shorten their life span.

4.2.4 When not to work on veteran trees

In many situations the right decision is not to carry out any work on the tree at all. This is especially likely to be the case where veteran trees are naturally occurring within semi-natural woodland.

Many veteran trees in other situations do not require any surgery work. If they are stable and in good condition (which many are) there is no need to do anything. This does not mean that a tree can be forgotten; it will need checking on a regular basis to make sure that the situation has not changed.

Working with veteran trees requires long-term vision, the temptation to work on them just to demonstrate that they are being managed and to show results quickly should be avoided.

4.3 How to carry out tree surgery on old trees and maximise the chances of success

4.3.1 Tree surgery on veteran trees

Once you have made the decision to cut a veteran tree it is necessary to look at the details of how and when to do this.

Only a few years ago the chance of a veteran tree surviving cutting was viewed as negligible. Since then a variety of veteran trees have been cut in different situations with mixed results. Some of the work has undoubtedly been successful, such that it is starting to become possible to give guidelines as to what is *most likely* to work but it is not possible to give a prescription of what *will* work. This depends on so many variables, eg tree species, age, soil type, location and aspect, previous management, skill of the operator, environmental conditions and the incidence of subsequent stresses that the tree may face.

If you are looking at working on a population of old trees try out the techniques on a small number first to ensure that it will work. This is also better for the conservation value of the site too, ie never do something to the whole of the resource at once. Sites with many veteran trees are all the more valuable in conservation terms because of the aggregation of trees. Individuals are not expendable just because there are many of them.

Two broad-scale objectives can be defined:

- cutting old trees with a view to making them safe, or saving them from imminent collapse as a one off treatment (remedial work);
- cutting them with a view to getting back into a (semi) regular pollarding/coppicing routine (restoration pollarding/coppicing).

In some situations this distinction is a little blurred, but you should think about the long-term future management of the tree because it does sometimes have implications for the work being done.

The emphasis here is on restoration work, and maximising the future survival of the tree. Similar principles apply to remedial work as it is always important to maximise the chances of the tree surviving in the future if the situation allows.

4.3.2 Species of tree

Each tree species seems to respond in a different way. Appendix 4 gives as much detail as is currently available on the likely success of work done on different species of veteran trees. The table below presents a *very rough* rule for guidance. There is considerable variation between different situations so this table should not be taken too literally.

Species	Ease of Cutting		
	Tolerance to Pruning in a Veteran Tree	Creating Young Pollard	Initiating Pollarding on Mature or Post Mature Maidens
Willow	★★★	★★★	★★★
Plane	★★★	★★★	★★★
Lime	★★★	★★★	★★★
Apple/Pear	★★(★)	★★★	★★(★)
Hawthorn	★★(★)	★★★	★★(★)
Yew	★★(★)	★★★	★★
Hazel	★★(★)	★★(★)	★★(★)
Holly	★★(★)	★★(★)	★★
Hornbeam	★★	★★(★)	★★
Sycamore	★★	★★★	★★(★)
Poplars incl. Aspen	★★	★★★	★★(★)
Field Maple	★★	★★★	★★
Sweet Chestnut	★★	★★★	★★
Horse Chestnut	★★	★★★	★★
Alder	★★	★★★	★★
Oak Spp.	★★	★★★	★(★)
Sorbus Spp.	★(★)	★★(★)	★(★)
Ash	★(★)	★★(★)	★(★)
Birch Spp.	★(★)	★★	★
Prunus Spp.	★	★★★	★(★)
Beech	★	★	★
Scots Pine	(★)	(★)	-

★★★ Likely to respond well to cutting
★★ Likely to show a medium response to cutting
★ Likely to show a poor response to cutting

Brackets indicate that the response is variable (according to location, etc).

4.3.3 Time of year

It is difficult to give a good prescription but **the times definitely to avoid are spring**, when the leaves are just opening on the tree **and autumn** when they are being lost. At these times it is considerably more difficult for the tree to deal with the stress of heavy pruning. In Britain traditional cutting seems to have been done in the winter and probably the ideal time for cutting is January to March. Slightly less ideal is November to December and it is probably best to avoid cutting altogether in frosty weather. However, cutting for fodder from most trees must have been done in the summer. Mid-summer cutting has been shown to be successful in some cases. Probably, severe drought years are best avoided (though these may have been the very years when additional fodder was required). July and August are probably the best summer months to cut in. However, there are other reasons for not cutting then. For example, there may be birds nesting in the trees, herbivorous insects are abundant and it is difficult to see the shape of the tree in order to decide where to cut it. See also Lonsdale (1994) for a discussion of the relative merits of cutting at different times of the year.

The following calendar gives a rough indication of the best times to cut veteran trees:

Jan	Feb	Mar	Apr	May	Jun	Jul	Aug	Sep	Oct	Nov	Dec
★★★	★★★	★★★	X	X	X	★	★	X	X	★★	★★

X Not good, ★ Possible, ★★ Better, ★★★ Best
There is also some variation between species, see Appendix 4 for details.

4.3.4 Amount of crown to remove (Figure 17)

Leave some limbs intact (and remove at a later date if appropriate). This is essential on some species and desirable on others. Small living twigs and branches all round the bolling should always be left if they occur (except perhaps on willow and poplar). The number of branches left should depend on the species of tree and its likely response to cutting. If it is a species likely to make a poor response, more branches should be left. If it is more likely to show a good response, leaving more than one branch may result in excessive regrowth in those that are left and little in the way of new ones. Some authors have suggested leaving a distinct central stem to make future cutting easier, but in practice, this often produces a more difficult situation in the future than leaving some branches lower down. Retained branches ensure that there are living pathways, for nutrients/water etc, from the pollard head to the roots. It is best to have these distributed round the trunk of the tree in species less likely to show a positive response to cutting. Cut according to the form of the tree and bear in mind any future cutting of the tree that might be needed.

Figure 17. *Diagram to show the amount of crown to remove and the amount of light reaching the tree.*

ALL BRANCHES REMOVED. TOO MUCH LIGHT FOR SOME TREE SPECIES.

SOME BRANCHES REMOVED SOME RETAINED. ENOUGH LIGHT IS REACHING THOSE CUT.

TOO MANY BRANCHES RETAINED. NOT ENOUGH LIGHT REACHING THOSE CUT.

4.3.5 Light reaching the tree

The amount of light reaching the veteran is important. The structure of the surrounding vegetation should neither give excessive shading nor lead to extreme desiccation. It is impossible to give a prescription for this but the ideal light regime for an individual tree will depend on the species and its location. Even shade tolerant species such as beech need plenty of light to grow successfully after cutting but the more light demanding hornbeam in open parkland can suffer from excess sunlight and heat. Points to consider are:

- any surrounding trees; these should not be overshadowing the veteran and the canopies should not be overlapping, (but beware of opening up round a veteran too quickly, see section 5.3);
- the branches on the veteran itself. Consider retaining branches that shade the south side of the bolling (or the exposed side) to provide shade after cutting.

Don't forget that even on the same site, trees may have different aspects so desiccation may be a problem on a south facing slope but not elsewhere. Sun scorch of leaves after recent clearing, or substantial reduction, can be a problem but is usually not fatal. On occasions a veteran tree may be shaded by another veteran tree. In this situation it may be necessary to carry out reduction work on both trees at the same time to ensure adequate light levels.

4.3.6 Length of stub left

The branch collar (a ridge of bark where a branch joins the trunk) should be left, and on no account cut into (Figure 18). General arboricultural practice is to cut just above the bark ridge as the tree can then more easily recover. Clean cuts made close to the previous cut or trunk do not usually produce good regrowth. When cutting above a side branch it is also important not to injure the branch bark ridge. Probably, the larger the diameter of the branch, the longer the stub that should be left (a longer stub is also more likely to have viable dormant buds than a shorter one). As a rough guide, leave 10 times the diameter of the branch above the bole of the tree.

Figure 18. *The position of the branch collar and where to make a cut.*

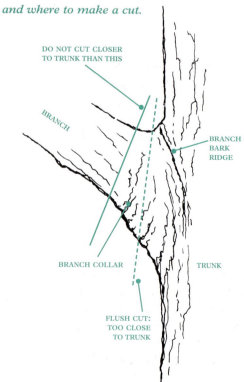

DO NOT CUT CLOSER TO TRUNK THAN THIS

BRANCH

BRANCH BARK RIDGE

BRANCH COLLAR

TRUNK

FLUSH CUT: TOO CLOSE TO TRUNK

Different experiences have provided conflicting information but the importance of leaving stubs does vary according to the species of tree and the age of the limb. Long stubs may encourage decay in the trunk but where the bark is thinner (away from the base of the branch) dormant buds are more likely to emerge so that leaving longer stubs may be justified. Larger diameter stubs are slower to callus over and tend to die back more.

It is also important to cut above the previous pollarding or re-cutting, preferably above healthy side branches (except if cutting in two stages see below).

4.3.7 Cutting in two (or more) stages

Depending upon the shape of the tree it is sometimes worth considering cutting it in two or more stages, several years apart (Figure 19). This has been done successfully with oak (White 1991a, 1996 and V. Forbes pers. comm.), white willow (Wheal 1998) and black poplar (L. Davies pers. comm.). In the first stage the upper branches are removed, a high proportion

being retained lower down. After an interval of one to five years (according to the species and growth following cutting) the second cut can bring the crown down to a lower level, retaining some of the new growth produced as a result of the first cut. This method may prove successful with other species but seems less likely with beech as new growth rarely occurs well below the point of cutting.

Figure 19. *Cutting a veteran tree in two stages.*

TREE BEFORE CUTTING.

TREE AFTER FIRST CUT.

TREE SEVERAL YEARS AFTER
FIRST CUT SHOWING GOOD
GROWTH.

TREE WITH LARGER BRANCHES
CUT BACK, SMALLER & MORE
RECENT BRANCHES RETAINED.

4.3.8 Type of cut

There are two schools of thought here: first that cambial regeneration is usually very poor, therefore there is no advantage in slanting cuts to increase its perimeter and thus it is best to cut at the easiest angle. The alternative is that a slanting cut is better because it sheds water and increases the chances of adventitious growth (albeit sometimes only slightly). It does not seem necessary to leave a clean cut. A jagged edge may encourage better growth from adventitious buds due to the increased amount of exposed cambium.

It probably does not matter what type of cut is left - experiment and find out what is best for your site. The visual appearance of cut surfaces may also influence your decision.

4.3.9 Cutting tool to use

It has been considered by some people that edge tools are better than saws, whereas others believe that there is no difference. There is also a school of thought that chainsaws are not good but there is no real evidence for this. The only experimental work to be carried out on tools was on sweet chestnut coppice. This showed that cuts with an axe produced more even growth than that with a chainsaw, where surges of growth result, although there was no overall difference in growth. Cuts with an axe showed new shoots nearer to the cut; with a saw there were more initial shoots but some died.

4.3.10 Weather conditions at the time of cutting

Avoid cutting in drought years (or the following year if it is very severe). Be careful when cutting trees that are in frost hollows, try to cut during a warm spell or at the end of the winter.

4.3.11 Good balance

Ensure that the shape of the tree after cutting is not unbalanced (Figure 20). Be careful though not to destroy an existing asymmetrical shape that is adapted to, for example, strong winds. This aspect is covered in detail by Mattheck & Breloer (1994).

Figure 20. *A tree well balanced after cutting and one that is unbalanced.*

TREE BEFORE RESTORATION WORK.   TREE WELL BALANCED FOLLOWING RESTORATION.   TREE NOT WELL BALANCED FOLLOWING RESTORATION.

4.3.12 Regional differences

Trees in the more humid west are less likely to suffer from desiccation than those in the east. Regrowth tends to be more vigorous in westerly districts and in mild wet areas such as the Lake District. Less responsive species tend to grow better following tree surgery here than in the south-east.

4.3.13 Growth of lower branches

In grazed areas, lower branches will be browsed and kept back. Where livestock or deer are not present some trimming of growth low down on the trunk may be necessary. Excessive branches lower down may divert energy from those at the top of the tree.

4.3.14 Age of tree and length of time since any previous tree surgery

The younger the tree and the less time that has elapsed since any last major cut the more likely the cutting is to succeed well (within species).

4.3.15 Trees with burrs

Trees may respond better to cutting if they have visible burrs or dormant buds. However, it can be difficult to tell and the differences are probably at least partly genetic.

When assessing a veteran tree with a view to carrying out tree surgery consider the following points:

• Which branches are alive and which dead? (ie when left, which will continue growing?)

• What type of growth pattern does it show? (Lots of epicormic or none? Any previous growth from dormant buds?)

• What is the growth form of the tree? (Are there obvious places to cut back to? Are there good branches to leave?)

• How well balanced is the tree? (If a branch is removed will the tree become unbalanced?)

A few general points on cutting veteran trees are worth stressing:

• For all species except willow and poplar, near 100% success rate cannot be achieved if all the crown is removed.

• The more 'difficult' a species is the more important it is to retain much of the existing canopy.

• For all species it is worth retaining any small or young growth around the bolling.

• Bear in mind that removal of the entire crown can cause excess drying of the bolling, especially if the tree is in open conditions.

• It should not be thought that because branches have been left, the tree is not properly pollarded. The main purpose of the work is to extend the life of the tree. This is far more important than details of the terminology. Retaining branches can be a short-term measure and after a few years the tree can often be re-shapes if necessary. It is also likely that some species never had the entire crown removed in any case.

• Wait for a few years to assess success. Don't assume that growth in the first year means that a technique has worked.

• Whatever the species don't cut all the trees at the same time, even if you know it will work!

Practices which are not recommended include:

• Cutting entire populations of old trees at the same time.

• Thinning the resulting growth. This can be done but the trees usually self thin (ie branches that do not have enough light die anyway) and unless resources are available it is not usually necessary.

• Sealing the cut with wound dressing. This does not help the tree to recover from the cutting. (Though it may help to reduce excessive drying.)

• Cutting every year. This is sometimes recommended for street trees especially in the USA but is *not* recommended for old trees on sites of importance for nature conservation unless it is continuing an existing practice.

Mulching of veteran trees

There is some debate about whether mulching veteran trees, by putting a deep layer of leaves or woodchips around the base, is beneficial or not. It is likely that this was an historic management technique, at least in Blean woods, Kent. When the Bishop traditionally visited the wood after September the leaf litter was swept away for him to walk and the leaves used to mulch the pollards (D. Maylam pers. comm.). More recently the Major Oak at Sherwood (see Figure 48) has been mulched and is showing increased vigour as a result. However, there is a concern that a thick layer of organic matter may encourage increased growth of roots close to the surface that may then be more susceptible to desiccation in dry periods.

> ### *Recording*
>
> *Make sure that you record what you do with your veteran trees. Photographs are a minimum requirement but more detailed descriptions of what was done and how the trees responded are more useful for future management. Consider also doing ring counts of any branches removed. See also chapter 12.*

4.4 Comments specific to pollarding

4.4.1 Cutting veteran pollards that are still in a regular cycle of cutting

There are some pollards that are still cut on a more or less regular basis and cannot be considered as lapsed pollards. These tend to be either urban trees (lime or plane that are responsive and easy trees to deal with) or trees in agricultural situations where the farm workers have continued to cut them. In addition, there is at least one place (Hatch Park in Kent) where an extensive population of veteran pollards has been cut continuously since the middle ages with no break.

In situations where there has been no lapse in the pollarding cycle the best course to take is to continue with whatever has been done in the past. Often this seems to be to remove the whole crown in species such as ash and oak. It is likely that trees in a regular cycle of cutting are more able to grow following removal of the whole crown than are lapsed pollards. It is doubtful, however, if beech pollards ever had the whole crown removed.

> ### *Pollards in farmland*
>
> *Goswold Hall in Suffolk is a typical arable farm of 150 ha. Around the edges of the fields, in the hedges and by the green lanes there are 15 ash pollards, plenty of oak pollards and two field maple pollards; prior to Dutch elm disease there were over 100 elm pollards. The owners took over in 1937 and most pollards have been managed continually since the (Figure 21). A few oaks were left to grow out, as prior to 1937 they had not been cut regularly and some had not been pollarded for over 100 years.*
>
> *In 1937 there were 10 men working on the farm and each man was given a hedge pollard, or a stow, usually elm, maple or a smaller oak, as a perk in the winter. They pollarded the tree in their own time and used (or sold) the wood for firewood. Stowing a tree (pollarding) was done on a five to eight year cycle by climbing the tree or a ladder and using a heavy billhook with a 2′ (60 cm) handle. The men considered that a tree should not be cut with a saw unless you had to because it would not shoot so quickly, 'it needed to be chopped'. If a saw was used there was a danger that the branches were cut too close to the trunk and then the tree might die. The bigger oaks were pollarded for fuel in the Hall and for structural barn timber. They were cut using saws with large double or triple teeth like a crosscut saw, with deep intermediate cuts to self clear but they were single handed with a wooden handle. Many had holes at the other end to which a second handle could be fitted quickly if necessary. The ash pollards were potentially cash crops, the trees were 'sold' to a contractor who pollarded the trees for making handles, especially for the local brush factory. There were also itinerant pollarders who pollarded the trees for a share of the crop.*
>
> *In recent years the trees have mostly been cut on a 12 - 15 - year cycle and up until the 1980s the cutting was done with bow saws; now chainsaws are used. Branches are cut within 2" (5 cm) of the bolling. A branch was once left on a veteran oak but it was removed after a few years as the tree was not growing well from the bolling. The pollards are now managed along with the coppiced hedges and grass headland under a Countryside Stewardship with £40 being paid for each pollard cut. The wood is used for the farmer's central heating systems.*
>
> **Figure 21.** *See colour plate page 85.*

4.4.2 Future cuts on lapsed pollards

When working on lapsed pollards it is important to consider the future. The purpose of tree surgery may be just to increase the short-term life expectancy of the tree. However, for some trees a cycle of (semi-)regular cuts can be considered, which will continue to increase the life of the tree. When cutting lapsed pollards it is important to think about the subsequent growth and any future cutting. Try to avoid creating pollards 'on top of' pollards if possible as this makes the trees difficult to manage in the future. Only a few lapsed pollards have been cut for a second time (apart from those cut in a series of stages, over a short time span) so experience of this is, as yet, very limited (Figure 22).

Figure 22. *See colour plate page 85.*

4.5 Who does the work on the veteran trees?

Carrying out tree surgery work on veteran trees is a dangerous occupation and should not be undertaken lightly. It is essential that *any* work done with a chainsaw is carried out by fully certificated operators wearing full personal protective clothing. In most cases it is also necessary to have certificates for using a chainsaw at height, and for tree climbing.

> ### *Managing* dead veterans
>
> *Dead trees may also need managing and should definitely be retained. They are tremendously valuable habitats but are often under threat of felling on the grounds of safety or tidiness. As with living veterans, try to do as little remedial work as necessary to make the tree safe, and avoid cutting and or removing features such as cavities. Sudden extensive clearing of younger woodland around dead stumps should be avoided as this can dramatically alter the light/moisture regime for the organisms inhabiting decaying wood.*

4.6 Managing populations of veteran trees

Populations of old trees are sometimes threatened when the land they are situated on changes owner or manager. The more changes, the more likely that the trees may be managed in an inappropriate manner. The fragmentation of land may create similar problems.

Many of the organisms associated with veteran trees require good populations of trees and are threatened by:

- the sites being too fragmented;
- trees being too widely scattered;
- a missing generation of trees. Thus there will be no new veterans in the near future and often no young trees at all;
- future generations of trees that are a different species from the old ones;
- no prospects for the long-term continuation of biological features characteristic of old trees, eg decay.

Populations of veteran trees should ideally consist of a large number of trees, in close proximity, with a good range of different ages (including future veterans). Many sites need active work to reach this ideal.

When contemplating the management of populations of old trees the following points need to be considered:

- A good age structure is achieved and/or maintained (ie there are abundant veterans and also a range of younger age classes to replace the veterans eventually).
- A good number of veteran trees are present (the exact optimum density is unknown but the more the better).
- Try to reduce site fragmentation (by promoting future veterans on land in between existing sites).
- Continuity through new generations of trees.
- Continuity of dead wood resource, both standing and fallen.
- Continuity of dead wood features, eg decay cavities, rot pools, sap runs.
- Pollard trees of intermediate age to close up the generation gap. Also encourage open grown trees, which develop larger butts and trunks and have more heart wood decay at an earlier age than woodland trees.

When managing populations of veteran trees it is necessary to think about what the situation will be like 100 or more years into the future. This of course cannot be predicted but managing for long-term continuity should be attempted.

Population dynamics of old trees

There are a few sites where trees have been surveyed and then the survey repeated after a time interval. There is scope for more detailed, repeated surveys and the establishment of a standard recording form for veteran trees (see chapter 12 and appendix 6) should help this.

Harding & Alexander (1993) showed that, for five parkland sites, the age structure of the trees was heavily weighted towards the older age classes. As a result proposals for planting schemes were drawn up for three of the sites. Losses of trees from three of the sites between 1976/79 and 1989/90 were generally slight despite strong winter gales in this period.

The trees at Duncombe Park in North Yorkshire were surveyed and mapped in 1986 (Clayden 1996). In 1994 1200 open grown trees were re-surveyed. Even trees of small girth (2-20 cm) were included. Losses ranged from 0% (lime) to 8% (ash) over the eight-year period. Losses were not necessarily in the oldest age groups. Higher percentage losses were recorded for wych elm (Dutch elm disease), holly and cherry. There was concern that for some species (ash, beech and field maple) recruitment may not be enough to ensure a continuous range of older trees.

Veteran pollard beech trees were first accurately recorded at Burnham Beeches in 1931 when there were 1795 trees. In 1956 the numbers had declined to 1300 and by 1991 there were fewer than 540 (Read, Frater & Turney 1991). The decline has added impetus to managing the veteran trees and creating a new generation of pollards.

Further reading: Countryside Commission (1998), Edlin (1971), Harding & Alexander (1993), Le Sueur (1931, 1934), Mitchell (1989), Read (1991, 1996) and see Appendix 4.

VETERAN TREES
I N I T I A T I V E

| Chapter 5 | Management of land around Veteran Trees |

5.1 Introduction

Trees do not exist in isolation; the environment around them can be crucial to their welfare and the landscape they are part of can be just as or even more important. The previous section has dealt with the management of veteran trees themselves. In many situations though, it is not the trees that need work done on them as much as the land surrounding them. Conditions in the soil will affect their roots and growth. What is growing on the surrounding land can compete with the tree for water, light and nutrients or present a fire hazard. Outside influences can affect the health of the trees and the organisms found on them.

Ancient trees are part of a landscape, either formal and designed or informal and evolved. It is important to consider what role the trees play within such landscapes as this will have implications for how the land around the trees and the trees themselves are managed.

5.2 Open land

5.2.1 Grassland

5.2.1.1 Grassland management (for full details see Crofts & Jefferson 1999)

The most sympathetic type of grassland to surround veteran trees (though not commonly found) is unimproved. Unimproved grassland should be managed with great care following the guidelines in Crofts & Jefferson 1999; this management will also be favourable to veteran trees. In general, the 'improvement' of grassland through reseeding or the application of fertilisers can be detrimental both to veteran trees and to grassland. For example, fertilisers and herbicides have had an adverse effect on the lichens and fungi at Moccas Park, Herefordshire, and the application of lime can cause problems for the fungal communities. If the grassland has already been improved and lime must be applied, it is better as coarse-ground limestone than agricultural lime as there is less drift. The ideal management is not to apply any substances although light applications of farmyard manure to improved grassland are acceptable in some situations. No inorganic fertilisers should be applied and no ploughing or reseeding carried out. Note that trees in fertilised pastures may look healthy but in times of stress they may decline quickly. The 'improvement' of pasture leads to a loss in variety and abundance of plants and invertebrates and therefore a loss of food for bats, birds, etc that may depend on the veteran trees for their resting sites. Hay cutting (the traditional management of grassland under pollards in other European countries) can be compatible with veteran trees though it is better if the grassland is not improved.

The control of problem species, such as ragwort, thistles and docks, within grazed systems needs to be considered carefully. There are methods that are sympathetic to ancient trees. For example, cutting of thistles, spot spraying, and use of some modern weed wipes that are relatively specific (see Bacon 1994). There are also suitable mechanical weed pullers available now (Bacon & Overbury 1998).

Effects of grassland improvement on veteran trees themselves

- Inorganic fertilisers disrupt mycorrhizal fungi, and the trees are then more susceptible to stress.
- Lime may reduce the species diversity of fungi (including mycorrhizal species).
- Farmyard manure, slurry, fertilisers and lime may be sprayed onto the trunks of the trees; an excess is toxic to fungi.

- Ploughing damages the roots of the trees and mycorrhizal fungi.
- Rolling compacts the roots (see also section 5.2.2).

5.2.1.2 Grazing

In many situations where veteran trees exist, grazing was an essential part of the management system. There are two quite different types of grazing that can be considered, extensive and intensive.

The benefits of grazing in wood-pasture are that:

- **It prevents mass regeneration of trees and shrubs, thereby reducing competition between trees for light and creating more varied growth forms of trees and a greater variety of woodland structures. This results in:**
  - A greater diversity of habitat structures, allowing other groups of organisms to survive.
  - Increased light levels that are beneficial to epiphytic lichens.
  - Many insects are sun-lovers as adults and will be lost under shady conditions.
  - Densely shaded trunks have cooler interiors and are less good for larval development.
  - Many wood-decaying fungi appear to fruit less frequently when the trunk is shaded.
- **It is the traditional form of management on many sites and, from an historic point of view, is an essential part of the system.**

Extensive grazing

Animals are able to roam over large areas to forage and the stocking density is relatively low. This type of grazing must have existed in the wildwoods, which naturally contained high densities of veteran trees. The grazing would have been by indigenous cattle, deer, bison and boar. Such grazing restricted, but did not stop, tree regeneration and had the effect of creating structural diversity in the ground flora and shrub layer (Figure 23). Today, woodlands with veteran trees benefit from light grazing, which increases the structural diversity and benefits a range of organisms (Mitchell & Kirby, 1990).

Figure 23. *See colour plate page 86.*

Intensive grazing

This type of grazing involves a higher stocking density and the land under the trees is dominated by plants that are not woodland species (Figure 24). The animals are often given supplementary feed, which is detrimental, to the ground flora and in many cases the grassland has been, or is under pressure to be, improved. Often the areas the animals range over are relatively small. Intensive grazing occurs in some types of wood-pasture (where the old trees were usually pollards), especially in parks where deer (or in some instances cattle) were grazed.

If a wood pasture or parkland has been grazed more or less without a break, it is likely that the most suitable management in both historic and biological terms is to continue with this. Changing the management regime in such circumstances should be carefully thought out before implementing. If there has been a long lapse in grazing, the situation should also be looked at more carefully, especially from a biological point of view, before deciding on what form of management to reinstate. (See also section 5.3 for a discussion of the consequences of opening up woodland from around veteran trees).

Figure 24. *See colour plate page 86.*

VETERAN TREES
I N I T I A T I V E

Disadvantages of grazing

There are disadvantages of grazing too, especially where it is intensive or the land is overgrazed:

- **Lack of tree regeneration.** On sites where grazing has been continuous there is often a generation of old trees with few, if any, younger ones to form the ancient trees of the future. This situation needs addressing to prevent the loss of biological interest on the site and loss of landscape appeal and historic value. There are several ways to overcome this (see chapter 8).
- **Activities relating to the grazing of animals that damage the veteran trees.** Trees in grazed areas will always develop a browse line, the height of which depends on the animals present. In a healthy mature tree this does not cause too much damage, although the trees are unable to grow branches that reach the ground. This prevents layering and may make the branches more prone to snapping as they cannot be supported from below (Lonsdale 1999a). The grazing of animals may be directly detrimental to the trees in the following ways:
  - Animals may chew the bark of the trees. In extreme situations they may ring bark them.
  - Animals may use the tree for shelter, causing trampling round it and damage to the roots. Where the animals dung and urinate, the nutrient levels rise and the high nitrogen levels are detrimental to the mycorrhizal fungi.
  - Vehicles used to feed and water animals may pass too close to old trees, causing the ground to be churned up.
  - Fodder, watering and mineral lick sites placed too close to the trees attract greater attention.
- Where the animals are fed supplementary feeds the dung enriches the grassland/heathland/trees. It may also introduce new species and genetic variability via seeds, which may confuse the true distribution and status of species.
- Animals bring other chemicals into the wood-pasture system. Domestic stock are treated, particularly for intestinal worms, using a variety of chemicals. Some of the substances used are not specific to internal parasites and may be long lasting. An example is the widely used wormer with Avermectin as the active ingredient. The action of this has been shown to have an effect on a range of invertebrates including those that break down animal dung. Insectivorous species, especially bats and birds, may consequently suffer from lack of prey. The situation is of special concern with regard to cattle because a bolus system is often used which releases the wormer over long periods of time. As a general principle Avermectins should not be used on any site with conservation interest. Other types of wormers may be just as detrimental. For more information see JNCC (undated), Cooke (1997) and English Nature (1994). See also later in this section.
- Overgrazing, especially in the winter months, can cause poaching and this leads to infestations of tall weeds such as thistles, nettles, ragwort and docks.

Solutions to problems caused by grazing

There are several solutions to these problems. The best one will depend largely on the situation and the money available for additional management work. They include:
- Reduce the stocking density or alter the grazing regime so that the animals do not bark the trees (but for husbandry reasons avoid single animals in an area). Bark stripping is more prevalent in the winter but can be due to boredom as well as a nutritional need for the bark.
- Experiment with mineral supplements (eg Uniblock produced by Dodson and Horrell) to provide the minerals that horses might otherwise obtain from tree bark; other supplements have been shown to change the diets of animals such as sheep. Be sure to place them well away from the trees.

- Fence off the veteran trees so that the animals do not have access to them. The fence should keep the animals beyond the extent of the canopy. This solution is usually practicable only if there is a small number of trees involved, and it can lead to the growth of competing vegetation.
- Ensure that there is nothing to attract the animals to stand under the trees. Feed and water them in a different place.
- Provide alternative shelter, with watering, etc to attract them preferentially.
- Ensure that vehicles do not use routes under veteran trees. If necessary move gateways.
- Try out other ideas to deter the animals, eg pile rocks round the bases of the trees to stop them from getting too close or chestnut paling wrapped round the tree (not as a fence). Be sure to check that they work (sheep can climb rocks and continue chewing the bark higher up!) and that they do not create other problems such as compaction or creating a humid day time resting place for slugs that then browse epiphytic lichens at night!

Solutions to the problems caused by wormers

- Don't use any wormer on site (ideal solution). Worm animals when they are grazing or housed elsewhere. Ensure that they are kept off the site for long enough for all traces of Avermectin, and other products, to have passed through (this will vary with the type of application). Avoid broad spectrum wormers and slow release bolus applications.
- Use wormers that are more specific to internal parasites. English Nature, JNCC and FWAG can provide advice on suitable alternatives.
- Be careful with new products, ensure that they are compatible with any nature conservation interest and seek independent advice.
- Consider carefully the use of other chemicals on the site. This does not mean that the health of the animals should be compromised. However, some chemical treatments for fly strike in sheep, for example, are 'better' environmentally than others.
- It may be worthwhile considering the guidelines of the Soil Association with regard to animal husbandry and the use of chemicals and medication, even if the aim is not to produce organic products.

Tree regeneration in grazed areas

In the New Forest, Sanderson (1996a) has shown that even palatable tree species can regenerate in grazed areas and this is also true in Hatfield Forest. Even in periods of heavy grazing thorny bushes, including holly, and dense bracken provide enough cover to allow young trees to become established. Regeneration requires extensive grazing systems that allow a full range of natural processes to occur, including shelter or respite from grazing.

Reintroducing grazing

The reintroduction of grazing on veteran tree sites that have not been grazed for a long period is starting to gain momentum. In these situations a site specific feasibility study is probably the best way forward. The choice of grazing regimes is likely to depend to a large degree on the wildlife interest of the site. In the majority of situations, however, a low density grazing regime is probably the most suitable. Achieving the right stock type and level at the right time is very difficult and may vary widely according to the type of land and also from year to year. When considering reintroducing grazing to a site it is important to consider the welfare of the animals, the work involved and financial implications. When considering reintroducing grazing to a site with an ancient monument make sure that the fencing and type of animals used are appropriate to the site and acceptable to the archaeologists.

Further reading: Bacon (1994), Crofts & Jefferson (1999), Lewis *et al.* (1997), Mitchell & Kirby (1990).

5.2.2 Bracken and heathland

The commons and wood pasture on which many veteran trees stand were not normally prime grazing or arable land. One of the contributing factors to the survival of many veteran trees was that the land around them was too poor (eg too low in nutrients, steep or rocky) for growing crops. In some areas heathland developed on the poor soils and where today, through lack of grazing and trampling, they have become dominated by bracken this can cause problems. Bracken has also become dominant in other vegetation types. The effect of the bracken fronds is very similar to dense summer woodland cover with few other plant species, including tree seedlings, able to get enough light. Bracken is also allelopathic, producing its own substance that inhibits the growth of other plants. The result is areas lacking in other tree species around the veterans. The dried fronds in bracken-dominated areas build up into a deep litter layer or thatch that early in the year can be a severe fire hazard. Fires can travel under the ground in the leaf litter and when this happens the old hollow trees act as excellent chimneys, fuelling the fire with air, which destroys the trees. For areas of heather-dominated land under veteran trees fire can be a problem too.

On sites with ancient trees and abundant bracken, some form of fire protection should be seriously considered. Ideally fire breaks should be made and kept open so that blocks of woodland are isolated from each other in case of fire and an area round each tree should also be kept bracken free, at least as far out as the canopy. In order to create such areas effectively, fallen dead wood may need to the moved (see the section 5.4 for details). However, this is not as easy as it sounds as bracken is a difficult plant to control. Health and safety considerations need be taken into account with some of the following treatments and an added problem is that bracken spores are carcinogenic.

5.2.2.1 Methods to consider for controlling bracken to create fire breaks or to reduce the fire risk on the site

- **Spray or weed wipe with Asulam.** When used properly this herbicide is very effective against bracken, but rarely eradicates with one application. Spot treatment of any regrowth is necessary in subsequent years. Prior to spraying, the ground vegetation should be surveyed because Asulam does kill *all* ferns and some other plants, eg sheep sorrel and some bryophytes. Surfactants, if added to the spray, may affect invertebrates. On many sites it is preferable not to use chemicals; however, using more modern methods of application the bracken can be targeted with minimal amounts of herbicide reaching other plants. Spraying is the best form of application, weed wiping should be carefully considered before use. The best time of year to spray is when the fronds are fully unfurled but have not begun to die back, usually between mid - July and the end of August.

> **Bracken control at Ashtead Common**
>
> *Fires present a significant risk to the veteran oak pollards at Ashtead Common. In 1990 there were over 100 standing charred pollards as a result of a large fire (Figure 24a, see page 86). The risk has been reduced in recent years by creating fire breaks, primarily by chemical control of the bracken. Initially, all dead wood was removed from the fire breaks, into large piles nearby. Although not an ideal situation, it was necessary to allow the safe access of machinery and ensure that the fire breaks would be effective. Bracken is then sprayed annually with ASULOX by one person on foot, using a high pressure lance with a flood jet and a tractor-mounted spray tank driven by a second person. Small areas are sprayed using a knapsack sprayer. The herbicide provides an effective control rate of around 95% which is enhanced by mixing ASULOX with an oil derived from oil seed rape known as CODACIDE. This ensures that the chemical is water-resistant and improves adhesion to the fronds. Thereafter fire breaks are maintained by an annual mechanical grass cutting programme.*

Figure 24a. *See colour plate page 86.*

- **Cut.** Cutting two to three times per year will reduce bracken cover but this will need to be repeated in subsequent years as bracken re-establishes itself quickly. The first cut should be made when the fronds have just started to unroll (usually late April/early May). The second cut should be made about a month later when the fronds reach a similar height again. In this way the plant puts maximum energy into growing and loses it all when the fronds are cut. If a third cut is possible this can be done at any time when the fronds are uncurled, but cutting for the first time in a year after July is not worth doing. One drawback of cutting is that the ideal time to cut is also when ground-nesting birds will be most vulnerable. Cut, dried and baled fronds can be used as livestock bedding.
- **Bruising.** Small rollers that crush the fronds are available. Rolling twice, the second pass at right angles to the first, improves the effect. When rolled in July on a hot day the bracken bleeds, which helps to exhaust it. A special 'bracken breaker' has been developed that is especially good at breaking the frond stipes and encouraging 'bleeding'. Rolling seems to reduce the vigour of the bracken by at least 50% and it is possible that continued rolling may result in eradication. Since rolling is best carried out later in the year than cutting fewer ground-nesting birds are likely to be affected but the later breeding nightjar may be vulnerable. Rolling may also be detrimental to anthills, reptiles and rare or solitary plants, and can break up any dead wood on the ground. The rollers designed for bracken may also adversely affect wavy hair-grass but ericaceous species seem to be able to withstand the treatment as long as they are not old or straggly. Heavier rollers may affect plants such as gorse and broom and also trees and shrubs. The compaction of the vegetation produced by rolling can be an advantage in helping to reduce the fire risk and encouraging decomposition, although the small specialist 'bracken breakers' cause bruising with negligible compaction.
- **Pigs.** Pigs in the autumn months dig for and eat bracken rhizomes and this can help to control the growth of the plant. Pigs should not be fed solely on bracken as this causes them to develop thiamine deficiencies. The results of pigs digging can be quite devastating, if not carefully controlled, so this method is most often used on recently cleared land. A low density of pigs can also help in the control of bracken if used in association with cutting, or rolling. Rooting by pigs is beneficial in promoting the germination of some seeds but 'mob stocking' can cause similar problems to that of overstocking with other grazing animals (see section 5.2.1.2).
- **Removal of bracken litter.** If the bracken litter, or thatch, is very deep it can be removed and this may also be desirable if cutting or rolling are being carried out. Removing the build up of thatch can also help weaken the plant allowing possible frost damage but it may also suppress the growth of other plants. Litter can be removed by 'blowing' using strong leaf blowers, which causes less root damage to trees than scraping. It may be possible to sell bracken litter to the horticultural industry, as a mulch, and this has been successful in the Netherlands.
- **Grazing.** Livestock, especially large animals such as ponies or cattle, may trample the young bracken fronds, which can help to keep it in check once major control has been done physically or chemically.
- **Woodland.** Encouraging closed canopy woodland also helps to reduce bracken cover, though this may not be a desirable option if the veteran trees are unduly shaded by the surrounding woodland.

5.2.2.2 Disadvantages of controlling bracken

Bracken is an important plant for some organisms. For example, it can provide suitable conditions under the fronds for the growth of violets, which in turn can provide an ideal habitat for some threatened fritillary butterfly species. Other species that may benefit are bluebells, an endemic weevil (beetle) and some fly species. If there are such species of

conservation value on a site with ancient trees then the management should take this into account. Bracken can also provide valuable shade (and frost protection) for piles of dead wood but fire breaks may be needed round these if they are large. In any case, bracken should not be viewed as a species to eradicate, since it is native. It may just need to be kept in check.

Further reading: Burgess & Evans (1989), Butterfly Conservation (1998), Crofts & Jefferson (1999), Forbes & Warnock (1996), Lewis & Shepherd (1996).

5.2.3 Cultivated ground

Ancient trees are also found within land that is cultivated. Many of those in the middle of arable fields have died or been felled but some manage to survive. In addition, those on the boundaries of fields are affected by cultivated land. Much parkland has also been ploughed and converted to arable farmland. Ideally, arable farming should not be carried out in areas with veteran trees. However, where it is unavoidable, steps should be taken to provide the best possible conditions for the trees.

A vital part of the tree, the roots, is out of sight, under the ground. Deep ploughing can be extremely detrimental as it destroys tree roots. Work nearer the surface, especially that leading to compaction can be equally damaging as some trees have abundant roots at quite shallow depths (Figure 25). Even harrowing and rolling can cause problems in compacting the roots and causing mechanical damage, this is especially true of modern power harrows.

Ideally, no work should be done closer to the tree than 5 m outside the extent of the canopy, or a distance from the centre of the tree of 15 times the diameter of the trunk at breast height, whichever is the greater. This establishes a 'separation distance' or exclusion zone round the tree and gives it the best chance of long-term survival. It can be quite surprising how far away the roots of some of these trees extend, eg up to 50 m.

Figure 25. *See colour plate page 87.*

Intense cultivation brings more potential problems. Spraying can be very damaging to the tree and its fauna and flora. Fertilisers have a detrimental effect on the tree, the lichens and the mycorrhizal fungi. Fungicides also affect the vital mycorrhizal species that the tree needs in order to survive and they also affect the lichen flora. Insecticides may kill the specialist dead wood species and are most damaging when used between May and July in the vicinity of the tree or on any plants in flower. These chemicals should not be used anywhere near old trees to avoid drift affecting them; the minimum distance is 15 times the diameter of the tree (as for cultivation). **Ideally there should be an island of uncultivated land surrounding the tree** but it is important that this is of sufficient size and is not 'eroded' each year. Ploughing and spraying right up to the trunk is extremely harmful. Veteran trees on the edges of fields are vulnerable to cultivation too. Remember that the roots will extend into the field, even if the trunk is in a hedge.

5.2.4 Amenity land

Veteran trees are also found in amenity areas. These range from private golf courses to school playgrounds and public parks. The same principles apply here as in other grassland or cultivated areas. Grassland tends to be very intensely managed in some of these situations, but the use of chemicals should be avoided close to the trees. An additional threat is the regular mowing and strimming, which if not carefully done round the bases of veteran trees can easily cause damage to the bark, exposing vital tissue. Wherever possible, try to create a 'nature area' of rough grass around the trees to alleviate these problems.

Trees in prominent positions are frequently used as locations for notice boards and signs. It is far better to use purpose-made posts but if trees must be used in this way, tie notices on, do not nail them. Make sure that they are removed before the tree grows into them. Old, neglected fences that are attached to trees cause a similar problem. Attaching fences to veteran trees should be avoided.

A frequent threat to veteran trees in amenity areas is compaction. Car parks, picnic sites and building development may reduce the amount of water a tree can obtain and lead to its rapid decline and eventual death. If a veteran tree is a valuable source of shade, consider providing alternative facilities nearby and relocate car parks, picnic benches, etc. Wherever people or property are in close proximity to a veteran tree the tree becomes a potential threat so the situation is best avoided.

5.3 Surrounding woodland

5.3.1 Veteran trees in woodland

There are three broad types of woodland which are found surrounding veteran trees:

- high forest woodland that has developed naturally (usually ancient woodland) which contains old trees, eg a particularly old specimen of an oak within oak woodland;
- naturally regenerated woodland (usually broadleaved) which has developed around veteran trees, for example on commons or wood pasture, in which the ancient trees are found;
- planted woodland (usually coniferous) where a commercial crop has been grown, the ancient trees being left at the time of planting.

The first of these three categories is the nearest to a wildwood but since British woodlands have been so actively managed in the past, all three situations potentially create problems for veteran trees. In the second and third categories especially, if the young trees start to grow over the top of the veteran, not enough light will reach the veteran tree. The natural ageing process and effects of any tree surgery are likely to reduce the crown of the veteran such that shading by younger trees has an even greater impact. Understorey bushes, such as holly or rhododendron, may compete with the veteran trees for water and nutrients (though in terms of competition for water, woodland may be less of a problem than grassland or bracken).

5.3.2 Removal of competing woody vegetation

As a consequence of these factors, on sites where previously open grown trees are now surrounded by others, some opening up may be necessary (releasing the trees). However, even this is not as straight forward as it might appear. The sudden exposure of a tree that has been shaded for many years can cause problems, for example:
- The leaves may be vulnerable to sun scorch.
- The tree itself may suffer increased transpiration rates and, with a root system that developed in a woodland situation, it becomes more susceptible to drought.
- Desiccation of the bark may stress the tree and cause cracking of hollow trunks through drying out in different places.
- General drying out of tree and increased exposure may be detrimental to the organisms associated with it, eg mosses and invertebrates.

These problems are inherent in any veteran tree canopy reduction but are considerably more pronounced when surrounding woodland is cleared away. The sudden opening up of old trees may also be more of a problem where conifers are involved because they cast dense year-round shade and create a cooler moister microclimate.

If you do need to open up around veteran trees, make sure that the species of interest associated with the site (and old trees) are known, and assess the impact that opening up of the woodland will have on them.

Some points to take account of are:

- Clear around the trees a year (or more) before doing any remedial cutting on the veteran trees (but be careful that the veteran tree is not exposed to greater winds).
- Clear round the trees in stages over a period of years. These stages may take five years or even ten. Be particularly careful if the veterans are in commercial forests in case all the surrounding trees are felled in one go.
- Consider clearing from the 'outside' in, ie leave trees closest to the veterans until last.

Clearing conifers from around veteran trees

Birklands is a woodland within Sherwood Forest (Nottinghamshire) containing some 500 veteran oak trees. The woodland has been planted up with commercial pines since 1935. In recent years work has started to restore 50 ha to oak woodland. Clear-felling of the plantation pines has been carried out on half of this area and crown thinning of the conifers on the other. Seven years or more later the result indicated that more veteran trees survived in the crown thinned area than the clear-fell and some of them showed stronger crown regeneration than in the felled area. The clear-felling was considered to involve a greater risk of desiccation to the trees and the dead wood (Barwick 1996).

Similar conifer clearance work is being carried out at Windsor Forest (Berkshire) (Searle 1996), Castle Hill (Yorkshire), Croft Castle (Herefordshire) and Ethy Park (Cornwall). As at Birklands, the indications are that thinning of woodland surrounding the veterans is more successful than clear-felling (Figure 26).

Figure 26. *See colour plate page 87.*

- Selectively fell, leaving some shade trees, especially if species such as birch, which cast a dappled shade, are present.
- Leave shade trees on the south (or most exposed) side.
- Pollard surrounding trees (if appropriate) to bring down the height of the surrounding canopy.
- Consider ring barking (or other equivalent methods) on surrounding trees to decrease the shade over a period. Note that some species take a long time to die when ring barked.
- Be careful not to expose the trees to other problems such as spray drift and air pollution.
- Avoid substantial clearance work in drought periods.
- When selecting which trees to fell and which to leave, remember that damaged ones (in terms of commercial silviculture) are more valuable for nature conservation.
- Fortunately, this means that the higher value trees commercially can be removed.

Obtaining the correct balance of light and shade round old trees is a challenge. Achieving it will depend largely on the species of tree and site concerned.

Rhododendron and holly

These evergreen species can be detrimental to veteran trees when the bushes become large enough. Being shade tolerant they can grow very close to the trunk of the veteran and compete for water in dry years. They are also serious competitors for light when tall. As they cast a deep shade all year round, be careful that their sudden removal does not cause excess desiccation of the trunk.

Some additional points apply to veteran trees in commercial forests:

- According to the UK Forestry Standard (Forestry Authority 1998) individuals or groups of over mature, veteran or pollarded trees should be retained in both broadleaved and coniferous forests where it is reasonably safe to do so. Also, some (number unspecified) dead or dying trees should be retained in regeneration areas.
- The risk of pests damaging trees or timber when ancient broadleaf trees or dead wood is retained in commercial plantations is extremely small.
- Standard forestry practices carried out on commercial conifer crops in areas of ancient trees may have a detrimental impact on the saproxylic fauna. Under conifers there is a cool climate and restricted light levels. The nectar sources are reduced and dispersal may be inhibited.
- Selective felling should also aim to leave an uneven age structure of retained trees and some trees to form veterans of the future (especially in deciduous forests).

The management of land surrounding veteran pine trees in, for example, Scotland is not considered here.

Retaining trees within harvested forests to become veterans of the future

Ideally:
- Aim for 5 - 10 trees per hectare.
- Keep trees that will not be in the way or become hazards to the public in the future, will not become overtopped by crop trees and are close to areas with conservation interest, eg plentiful dead wood, glades.
- Encourage them to develop a full crown.
- Consider making pollards, if a full crown is not appropriate, but remember that they will need to be managed in the future.
- Select native, longer-lived species such as oak, ash and beech. Keep some others such as, willow, wild service and other fruit trees, which are valuable as nectar sources or have a distinct invertebrate fauna.

Bear in mind that the harvesting of the crop trees will have a large impact on retained trees, especially if the crop is coniferous.

Further reading: Alexander, Green & Key (1996), Barwick (1996), Crofts & Jefferson (1999), English Nature (1994), Forestry Authority (1998), Forestry Commission (1990), FWAG (1997), JNCC (undated), Key & Ball (1993), P. Kirby (1992), Lonsdale (1999), Sanderson (in prep.), Searle (1996), Wall (1996), Winter (1993).

5.3.3 Storm damage

Storm-damaged woodlands can contain large amounts of dead wood. In the short term this provides potentially good conditions for saproxylic species but over-zealous tidying up can result in a significant loss of dead wood habitat. A code for dealing with such woodland is given in English Nature (1994). After a storm event in a woodland, follow the guidance in section 5.4 together with the following additional points:

- Living but broken trees should be kept if possible. If necessary (ie in public areas) reduce any hazard without felling the tree completely. Try to retain a mixture of age groups.
- Trees where the crown has been blown off but the bole remains should be left if possible. Some species may regenerate and continue to live; others may not but will provide a valuable dead wood habitat while they decay. Reduce any potential hazards if necessary.
- Broken branches and stumps should be left and not sawn flush.
- Fallen mature timber should be left **in situ** as far as possible. Especially, try to retain pieces that are badly damaged or show signs of decay. If some has to be removed to a different position use the same criteria as in 5.4. Reduce pieces in size as little as possible and keep the branches as intact as possible.
- Leave some fallen trees with upturned root plates; these are beneficial for insects such as solitary bees and wasps.
- Do not leave the wood for some years and then come in to clear and destroy it; many organisms will be destroyed with it.
- Do not remove, treat or burn stumps.
- Leave the wood to regenerate naturally if possible. If planting is necessary then use a species composition appropriate to the area.
- Leave some of the open spaces/glades created by the storm.
- Attempting to stand up wind-thrown veteran trees is usually unlikely to succeed, though reducing the 'sail area' helps increase the chance of survival. Some species may continue growing horizontally, and some may naturally layer, in which case they should continue to be looked after as other veterans.

5.4 Management of dead wood

The management of dead wood is an important aspect of the management of ancient trees. While the survival of the trees themselves is not usually affected by what happens to their branches when on the ground, many of the organisms associated with veteran trees can be found in fallen dead wood. If all fallen branches and slightly decayed trees are removed from a woodland, it may be impoverished by the loss of more than 20% of its species. All types of dead wood are valuable but some are more so than others, with each type (depending on its species, state of decay, size, etc) supporting its own assemblage of invertebrates. The decay rates of logs on the ground are very variable depending on the species and situation but as an example large oak logs in the USA are predicted to take over 170 years to disappear.

The conservation value of dead wood has received much greater attention in recent years but there is still plenty of room for improvement in terms of our management of it. The following principles of managing dead wood can be applied to almost any site, particularly woodlands. They are especially important on sites with large numbers of old trees and/or large numbers of saproxylic species.

5.4.1 Dead wood on the ground (Figure 27)

- **Size and shape.** Size matters and the bigger the better! The more the internal temperature and humidity of the dead wood is buffered from drying out and very high temperatures the greater the number of organisms it will support. The smaller the piece of timber the higher the surface area to volume ratio is, which causes greater temperature fluctuation and desiccation. Large diameter branches and tree trunks should be cut up as little as possible, preferably not at all - they should certainly not be cut into rings as these dry out quickly and in public areas are subject to disturbance. A range of 'natural' sizes is useful. They also look more natural if they have broken rather than having sawn ends.

Figure 27. *Diagram of an 'ideal' piece of dead wood.*

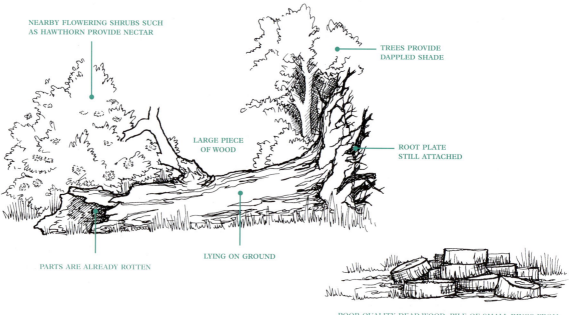

NEARBY FLOWERING SHRUBS SUCH AS HAWTHORN PROVIDE NECTAR

TREES PROVIDE DAPPLED SHADE

LARGE PIECE OF WOOD

ROOT PLATE STILL ATTACHED

LYING ON GROUND

PARTS ARE ALREADY ROTTEN

POOR QUALITY DEAD WOOD. PILE OF SMALL RINGS FROM A FELLED TREE LYING IN THE OPEN.

- **Position.** Ideally it should be left where it falls. If this is not possible, the ideal place to move it to is somewhere with dappled shade but:
  - Some organisms prefer wood in open conditions, eg jewel beetles, solitary bees and wasps re-using beetle holes and dead wood lichens.
  - Dead wood fallen into water should be left if possible. A different range of organisms live in it.
  - Dead wood in a range of different conditions is ideal.

Shade can also be encouraged, eg brambles/bracken can be left when they grow up. Freshly fallen dead wood is best in more open conditions and more decayed wood is better in shade. If felling a tree with the intention of leaving it as dead wood it can be felled into an open situation if it is likely to be shaded in the future by regeneration.

If it is possible to leave dead wood on only part of the site, choose that part with the highest potential value for saproxylic species (ie with dead wood of good quality and quantity and with good nectar sources).

- **When moving wood:**
  - Move as short a distance as possible.
  - Keep wood as intact as possible.
  - Move as soon after cutting/falling as possible.

- Move into partial shade/sun.
- Leave on the ground, not on top of other wood.
- Move near nectar sources (beneficial for many dead wood insects).
- Move adjacent to dead wood in a more advanced state of decay (to provide continuity of habitat).

- **Species of tree.** The wood from native species of tree especially those naturally occurring on the site is best. Introduced species can be valuable (eg sweet chestnut rots in a similar way to oak). Longer lived tree species tend to support a wider range of invertebrates. Broadleaves are more valuable than conifers except in Scotland (and perhaps a few other areas such as the Breckland).

- **Wood with decay.** Dead wood showing any sign of decay should always be left. It is more valuable for wood to decay from the inside out than from the outside in - decay derived from internal heartrot is likely to be far more valuable than that initiated from the bark or from the cut ends of sawn timber (Figure 28).

Figure 28. *See colour plate page 88.*

- **Burning** is a major source of damage to dead wood. Bonfires to burn brash (small branches cut from trees), or timber, kill living organisms, damage the soil structure and can physically damage living trees if they are lit too close. However, some organisms live in burnt wood. If a bonfire has been lit, do not start the next one with the charred wood of the previous one.

Figure 29. *The design of a 'Waterhouse' log pile.*

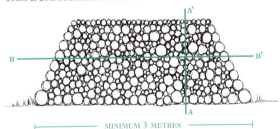

WOODPILE CONSISTING OF TIMBER FROM 10CM DIAMETER UP. MIX 10CM & 20CM DIAMETER FOR A GOOD PILE.

MINIMUM 3 METRES

1.5 METRES

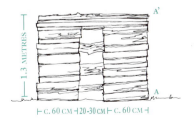

1.3 METRES

C. 60 CM   20-30 CM   C. 60 CM

- **Removal of wood.** Do not remove dead wood as firewood. If you have to remove it do not leave it on site to mature first. Make sure that any felled/cut is removed before the end of April or left permanently. If it has to be left and stored on site cover it with a polythene sheet. This speeds up the seasoning process and prevents invertebrates from laying eggs in it.

- **Habitat piles.** Tying small diameter twigs into tight bundles that retain moisture better can increase the value of them. They may then attract some specialised species.

Loose 'habitat piles' may be of value for vertebrates and a different range of invertebrates. Habitat piles made from larger diameter logs are better but the logs are always more valuable if on the ground (ie not on top of other logs). Piles of smaller logs are usually more valuable if lashed together or made into a 'Waterhouse' log pile (see Figure 29).

- **Root plates and stumps.** Leave the root plates of fallen trees as they fell unless they constitute a safety hazard. They can be good for nesting bees, wasps and birds and the holes left in the ground provide valuable damp patches/pools.

  Leave tree stumps in the ground. Consider cutting high stumps if this does not interfere with other management techniques.

- **Brash on grazed sites.** The retention of brash on heavily grazed sites can have a useful function in protecting tree regeneration. Stock can be prevented from getting caught in the dead branches by arranging them in such a way as to block access.

5.4.2 Standing and canopy dead wood.

Dead wood in the canopy of trees is different from that lying on the ground. Together with standing dead trees it forms an extremely valuable habitat (Figure 13). A variety of saproxylic species need timber to be still standing, either to maintain the moisture regime or aspect or because the 'search images' of the prospecting colonisers are geared only to habitats on standing trees. Broadleaved trees when they have died standing lose their twigs first, then small branches and bark. They tend to lose larger branches from the top downwards. As they offer little wind resistance they often remain standing for a long time and may take decades to decay. They do need periodic checking for stability. If necessary remove the branches and leave the trunk standing (creating a 'monolith'). Fallen trees can also be fixed in a vertical position, by strapping to living trees, to perpetuate this habitat type. Small dead elms should also be retained if possible.

> ### *Creating standing dead wood by ring barking live trees*
>
> *Ring barking is the removal of the bark and cambial layers all the way round a tree. This prevents the distribution of water and the products of photosynthesis and, over a period of time, kills the tree. It can happen accidentally, for example by grazing animals or grey squirrels, or intentionally. Ring barking can help create standing dead trees but they often decay from the outside, which is not quite as valuable as a natural decay column in the middle of a tree.*
>
> *When considering ring barking a tree, make sure that it is not in a position where it will constitute a hazard in future years. The ring of bark removed needs to be both wide and deep to be effective.*

5.4.3 Dead wood in living trees

The dead wood that is found in living trees is especially valuable for saproxylic organisms. The following guidelines should be followed in order to maximise the quantities of this rare habitat.

- Leave decayed wood inside hollow trees; be careful when carrying out remedial work.
- Leave dead limbs on trees.
- Be careful not to cut into cavities or damage them in any way.
- Leave dead bark.
- Keep commercially poor trees (they are usually the best for dead wood species).
- Try to retain trees that are due to be felled because they contain defects (they usually have interesting features), or at least retain the wood.

VETERAN TREES
I N I T I A T I V E

- Ensure that there is a continuous supply of dead wood on the site. Use short lived or fast growing species such as horse chestnut or birch to close up the generation gap if necessary and/or consider inducing rot in some trees.
- If branch removal is necessary, cut back only the weaker ends of the branches or cut part way through and knock them off to leave a natural shattered branch stub.

5.4.4 The quantity of dead wood

It is difficult to put a figure on the amount of dead wood required for saproxylic species. Broadly speaking the more the better and, ideally, all should be left. It has been estimated that one hectare of undisturbed woodland produces six tonnes of dead wood annually, equivalent to half the annual leaf fall, and that in the wildwood three to eight trees per hundred were standing dead. A suggested aim is for over 40 m³ of dead wood over 5 cm diameter per hectare and more than 50 standing dead trees, some over 40 cm in diameter. This is considered a good site by Kirby *et al.* (1998). **Quality is better than quantity** and the aim should be for a good representation of all successive stages in decay on the site.

There are various methods of estimating the amount of dead wood, both standing and fallen. That most widely used is outlined in Kirby *et al.* (1998) and given in Appendix 5. A simpler method is being developed by the Woodland Trust.

5.5 Summary of how to manage surrounding land for veteran trees

The key to managing the land around veteran trees is to remember that anything that is carried out to the land will also affect the trees. Aim for as little disturbance and input of substances as possible, especially close to the trees themselves. Extensive grazing is compatible with veteran tree management but intensive grazing and cultivation are not. Try to avoid any sudden changes of regime especially where these greatly alter light levels or affect the root systems. Wherever possible reduce chemicals or additives used on the land or livestock. Finally, avoid physical damage to the tree (especially the roots) either intentional or accidental.

Further reading: Alexander, Green & Key (1996), English Nature (1994), Ferris-Kaan *et al.* (1993), Fry & Lonsdale (1991), Fuller (1995), Green (1996c, 1997), Harding *et al.* (1988), Hodge & Peterken (1998), Key & Ball (1993), Kirby (1992), MacMillan (1988), Peterken (1996), Speight (1989), Spencer & Feest (1994), Watkins (1990), Winter (1993).

Chapter 6   Managing Veteran Trees of landscape and cultural interest

The land surrounding veteran trees may be very important for their survival, and appropriate management can enhance the conservation value of the tree or the site. The situation of the tree is also important for its contribution to the landscape itself.

6.1 Veteran trees in the wider landscape

6.1.1 Introduction

The over-riding importance of veteran trees in landscape terms lies in the trees themselves. While replacement is no substitute for the conservation of the veterans, detailed survey and careful re-planting programmes are needed to perpetuate their pattern within the landscape. The choice of species, source of stock, siting, planting and subsequent management are all vital, in order to provide continuity.

In different parts of Britain the 'Countryside' has a very different feel to it and the ancient trees often reflect this. For example, the flat landscape of the Vale of Aylesbury with its black poplars is very different from the undulating Lake District with field boundary pollards (Figure 30). This in turn has a very different character from the wooded nature of the Chilterns. It is important that this local distinctiveness is maintained and that cutting practices in terms of tree types and shapes continue the local traditions as far as possible.

Figure 30. *See colour plate page 88.*

Following the devastation of the elm population through Dutch elm disease many of the large trees remaining within hedgerows or along the edges of fields and roads are oak or ash trees as maidens or pollards. These are now especially vulnerable because of management for safety reasons along highways. Recent emphasis on planting and managing hedges should take into account the importance of leaving some trees to grow on. In the meantime, existing old trees need to be treated sympathetically to ensure their survival and that of the general aspect of the area they are found in.

Certain older landscape features also have historical associations with trees, many of which are now veterans, for example earth banks delimiting areas of former ownership or management. It is equally important that these features are retained and managed.

6.1.2 The management of veteran trees (and future veterans) in the landscape

Studying the characteristics of veteran trees in the local area can be valuable in perpetuating the character of the landscape. The following are some ideas to consider:

- Pollard (especially young trees) if there is a history of pollarding on the site or in the local area. If there is no historical reason for pollarding check that the historical value of the site is not compromised by doing so.
- Cuts on trees. In an area where trees were actively managed, flat and even cuts can look quite in keeping. If the overall atmosphere is to 'look natural' flat cuts can be quite intrusive. In this case try methods such as pulling or winching to make more natural looking breaks or roughen up the cuts after they have been made. This can be done simply by making coronet cuts, or V shapes, with a chainsaw in the cut surface (Figure 31).

Figure 31. *See colour plate page 88.*

- Be careful when planting new trees or encouraging regeneration. Make sure that they are in the most appropriate position, for example don't destroy vistas and do not create competition for existing veterans.
- When replacing trees (either by natural regeneration or planting) try to use the same species of local provenance and, if possible, trees that have very similar characteristics. Species that have been 'lost' from the site can also be used.
- The planting of exotic species, for other than timber production, is not necessarily a problem for nature conservation and may be desirable from the landscape point of view - to maintain traditional or planned planting patterns. Exotics should usually remain in the minority on sites with high nature conservation interests.

6.2 Veteran trees on ancient monuments

Where veteran trees are sited on an ancient monument their management must be carefully considered (Figure 32). While it is likely that disturbance by the tree's root structure has already occurred, further physical damage to the monument through collapse of the upper parts of the tree or lifting of the root plate (and underlying archaeological features) continues to be a potential threat.

In these cases it is important to limit the damage that may occur to the monument. Regular inspection is necessary and, where possible, reduction of the height/weight of the tree. If it is of suitable shape, encouragement of growth lower down, prior to cutting, may help to keep the tree alive without damaging the monument.

An additional factor to be considered in the management of veteran trees on ancient monuments is the depth and structure of the substrate. On ancient monuments this is frequently compacted, encouraging a shallower rooting pattern than is encountered elsewhere in the area.

Advice can be sought from the inspector of the appropriate heritage authority. If works are likely to affect an ancient monument, Schedule Monument consent may be necessary (see chapter 11).

Figure 32. *See colour plate page 89.*

6.3 Veteran trees in designed landscapes

When considering the future management of historic parks or gardens the starting principle should be to conserve, and where necessary repair, the surviving historic fabric. Veteran trees are as much a part of that historic fabric as the structures - from the main house or hunting lodge, to the park wall or park pale - which lie within the park. They should be valued equally as individual features and for their contribution to the wider parkland landscape. In general there should be a strong presumption in favour of keeping them and ensuring their future survival.

For those sites with public access, matters of health and safety can arise and must be given serious consideration (see chapter 10). Similarly, in more formal areas such as gardens, half dead trees and fallen timber might detract from the design or be viewed as evidence of neglect by the visiting public. In these cases a balance should be struck and where removal is considered necessary the timber should be re-located to a less formal area for creation of a habitat. Better still, in some situations this problem can be solved through education and explanation.

At a limited number of sites, the veteran trees will themselves form the basis of an overall landscape design. Usually this happens where the trees have been planted, or incorporated, as part of a late 17th or early 18th century layout that has not undergone the more usual reworking as a result of changing fashions. At such sites there might well be a debate about how to best manage the landscape for the benefit of present and, more particularly, future generations. This becomes more pertinent as the design loses its integrity and becomes gappy.

For example, avenues planted on new lines may result in the loss of much of the historic interest. Felling sections causes a temporary loss of historic fabric, as well as natural habitat, but provides continuity of historic design and new generations of trees.

As the above suggests, good management of veteran trees within a designed landscape needs to be considered in detail in order to maximise understanding and appreciation. Using site-based and documentary research, a comprehensive site plan can be drawn up against which the value of the landscape as a whole, and its individual parts (including veteran trees) can be assessed. Only by working through such a process can an informed decision be made about priorities for protecting veteran trees, the historic interest and valuable habitats.

Further Reading: English Heritage (1998), Phibbs (1991), Rackham (1986, 1988, 1989, 1990).

6.4 Veteran trees in a modern world

Veteran trees are often features in their own right, for example outside houses, pubs (Figure 33) or churches, on village greens and in prominent places. When managing trees in these situations the views of local residents and visitors must be considered. These types of tree are often unlikely to be felled because many people appreciate them. They are, however, vulnerable to concerns about their safety.

Figure 33. *See colour plate page 90.*

Veteran trees also occur in urban or suburban surroundings. It is unusual to find populations of ancient trees in such places (though they do occur) but individuals can be found in the most unusual situations. Sometimes it seems as if the threats are so great and the space the tree is in so confined that encouraging the attributes of old trees in such places is a foolhardy occupation. However, one of the chief management techniques used on old trees is quite applicable to urban situations and can help create veterans of the future: pollarding.

Some street trees such as plane, lime and sycamore respond very well to pollarding or heavy pruning. Although urban trees may be pollarded, this is not done for the same reasons as trees in rural settings. The aim is to keep the tree to a manageable size and any technique that achieves this can be used. In addition, regularly cut, responsive trees develop unusual appearances, which can add character to the neighbourhood (seen more regularly, for example, in France than Britain). If the work is done for practical or aesthetic reasons the result can be a tree with ancient features. There is no reason why characterful trees should not be found on city streets as well as in the countryside. Although street trees have been understudied in this respect, they may still offer valuable sites to wildlife.

> *An urban habitat can be a hostile environment for veteran trees, which can suffer from:*
>
> * excessive safety work, trimming and 'tidiness';
> * severing of roots caused by the digging of trenches for cables etc. Excavation work should not be carried out within a separation distance, extending away from the tree for 15 times the diameter of the trunk at breast height (ie 30 m for a tree of diameter 2 m). This should be regarded as a minimum;
> * run off from roads polluted with salt and trace elements from worn tyres;
> * tarmac, concrete or other unnatural substances right up to the trunk of the tree causing drought conditions;
> * excessive compaction round the roots;
> * vandalism and damage;
> * high nutrient levels from dog excreta.

It is important to realise that pollarding trees in this, as in any setting, needs to be done carefully by tree surgeons who show proven experience of working with veteran trees. Lopping the top off a tree that is too big for its setting is not the same as planned and careful pollarding. Poorly cut trees may be regarded as eyesores, incur the wrath of local residents and at worst be felled. In towns the future management of the trees is particularly important. Pollarding a tree once is not an appropriate method of dealing with it. A freshly made pollard needs to be cut at fairly frequent intervals (which can be as short as one to two years for street trees) so that the branches do not grow too big. If a gap in the cutting regime occurs, the task of dealing with the tree becomes substantially more difficult in the future. Today there are lapsed pollards on the streets of our towns that are in need of attention but are more likely to be felled and replaced. Pollarding some of these old trees might still be worth trying, otherwise the numbers of old plane trees, for example, will decrease dramatically in forthcoming years.

With planes and limes there is plenty of scope to create artistic 'patterns' through pollarding by cutting at higher points than previous cuts. Aesthetic proportions have even been suggested using 'golden mean proportions' where the stem of the tree should comprise 62% of the total height and the branches 38%. These dimensions produce an outline that is thought to be pleasing to the eye (Figure 34).

Figure 34. *An illustration of 'golden mean proportions'.*

Primary pollard

Secondary pollard

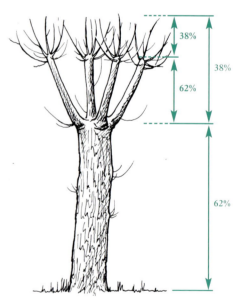

Further reading: Coder (1996), Mayhew (1993).

6.5 Educational opportunities

Veteran trees provide ideal educational opportunities for people to learn about their local environment and traditions. Collecting seeds and growing them on to plant out is a good starting point and enables local communities to 'connect' with their local ancient trees; this is the main focus of the Trees of Time and Place campaign. Activities promoted by organisations such as Common Ground also help to provide a focus and structure for such work and distributing information. Making children aware of the importance of trees, especially veterans, is a great investment for the future but adults should not be forgotten. The use of voluntary parish tree wardens helps to involve people and inform them about the importance of ancient trees and promotes tree recording as a contribution to the knowledge of trees in Britain.

Trees of Time and Place

This national project aims to increase awareness of the value of trees, especially among members of the public not normally involved with conservation issues. Headed by Chris Baines, a partnership of 80 bodies from corporate, public and voluntary sectors encourages individuals, school children and members of community groups to collect seeds and grow them. The seeds should ideally come from a specially chosen tree that has a cultural value, particularly one linked to a story or of personal interest. A pack, which emphasises choosing native tree species, has been produced with instructions on how to plant the seeds and care for the seedlings. The aim is to celebrate the millennium by planting the young trees back into the landscape to become feature trees of the future. Spin-offs from the project include celebration of trees through poetry and the 'Constituency Oaks', the planting of oak trees by Members of Parliament. Initially a three-year project, it is hoped that Trees of Time and Place will continue to provide encouragement and support for looking after the young trees once planted out.

| Chapter 7 | Management of Veteran Trees for other organisms |

7.1 Introduction

One of the reasons that veteran trees are so important is because of the range of other organisms that live on them or are associated with them. Indeed, many sites with populations of veteran trees have one or more statutory nature conservation designations for the species they support. For this reason it is essential to manage with other organisms in mind, not just the trees themselves. Many of these other species are important to the survival of the trees too, so they should be viewed as an integral part of the system. A number of the species associated with veteran trees are protected in their own right, via the Wildlife & Countryside Act, and many are listed in the Red Data Books or in the UK Biodiversity Action Plan and are therefore considered vulnerable or threatened. Ideally, the site or tree should be surveyed to find out which species are present and management can then be targeted. In reality detailed surveys are rarely achieved in the short term, but try to involve various experts early on if possible.

> *Indices of Ecological Continuity* have been drawn up for lichens and beetles based on the species that are more or less confined to old pasture-woodland and pollards. Similar indices may be developed for some other groups of organisms, but may not be possible for others, such as mosses, as there are too few species that can be used.

7.2 Managing for a range of organisms

Despite the variety of groups that species associated with veteran trees belong to, many benefit from very similar management practices. These practices are beneficial to the veteran trees themselves too and should be considered together with chapters 4 (management of the veteran tree) and 5 (land surrounding the veteran tree).

7.2.1 Management of the tree itself (Figure 35)

The majority of species will benefit from the following:
- Try to keep individual trees alive for as long as possible; live trees continually produce dead wood as well as leaves, and branches.
- Do the minimum amount of surgery necessary on a tree.
- Ensure that there is plenty of standing dead wood (including whole dead trees) and dead branches on old trees. Try not to remove the lower branches of trees, eg to allow vehicular access. If they are dying, due to shading from above, they might be used by some specialist insects.
- Never cut into cavities or holes, or drain them. To avoid this, test the depth of the cavity by using, eg a piece of flexible hose, inserted into the hole, and ensure that any cutting necessary does not go into it.
- Try to avoid damage to the lower parts of the tree trunk, including damage by grazing animals (rubbing or chewing etc). As well as harming the tree itself, such damage may be detrimental to other wildlife, for this is where lichens grow and there may also be cavities at ground level, which can be good for invertebrates.
- Don't tidy up (ie flush cut) rough ends to branches, the broken ends form egg-laying niches.
- Leave any dead wood in the canopy.
- Don't treat stumps or cut/damaged branches with sealant, fungicide or insecticide.
- Do not remove fungal fruiting bodies; it can be harmful for the fungus and also for any organisms living in it.
- Do not plough close to veteran trees, this damages the mycorrhizal fungi as well as the tree.

Figure 35. *An 'ideal' veteran tree for wildlife.*

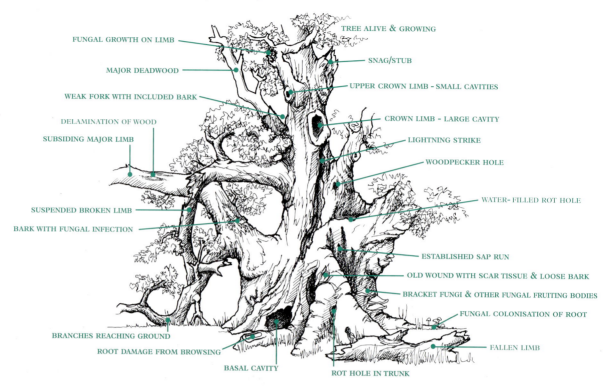

FUNGAL GROWTH ON LIMB

TREE ALIVE & GROWING

MAJOR DEADWOOD

SNAG/STUB

WEAK FORK WITH INCLUDED BARK

UPPER CROWN LIMB - SMALL CAVITIES

DELAMINATION OF WOOD

CROWN LIMB - LARGE CAVITY

SUBSIDING MAJOR LIMB

LIGHTNING STRIKE

WOODPECKER HOLE

WATER- FILLED ROT HOLE

SUSPENDED BROKEN LIMB

BARK WITH FUNGAL INFECTION

ESTABLISHED SAP RUN

OLD WOUND WITH SCAR TISSUE & LOOSE BARK

BRACKET FUNGI & OTHER FUNGAL FRUITING BODIES

FUNGAL COLONISATION OF ROOT

BRANCHES REACHING GROUND

ROOT DAMAGE FROM BROWSING

FALLEN LIMB

BASAL CAVITY

ROT HOLE IN TRUNK

7.2.2 Land surrounding a veteran tree (Figure 36)

- Ensure that there are plenty of holes, cracks and crevices, in other trees in the surrounding area eg for bats, birds and invertebrates.

- Leave abundant dead wood on the ground in a variety of sizes, shapes, positions and states of decay. Leave fallen dead trees as intact as possible.

- Poor or damaged trees are often those removed by foresters. Try to retain them if possible; they are often the best wildlife trees.

- Avoid using chemicals (herbicides, insecticides or fungicides) on the surrounding land (or the tree), keep the use of veterinary chemicals, especially wormers, on livestock to a minimum.

- If fertilisers have to be used, farmyard manure or pelleted versions are best. They should be applied on still days and kept at least 15 times the diameter of the trunk at breast height away from the trees and not allowed to splash onto the trunks.

- Encourage natural regeneration to ensure long-term continuity of trees. Try to encourage native trees and shrubs with a good population and age structure. This provides continuity of trees and suitable habitats for mycorrhizal fungi that require different age classes of each species of tree. The regeneration and planting of conspecific saplings near isolated veterans is important for this reason. On parkland sites with good lichen floras some younger exotic trees are worth encouraging if veteran specimens of the same species occur.

- If dead wood is in short supply, or will be in the future, (ie there is a generation gap) consider artificially creating suitable cavities and decay in younger trees (see sections 7.2.3 and 8.3.2).

- If there is no new generation of the same species of tree consider using other, more quickly growing, species to try to help close the gap as well as planting conspecifics.

> *Birch is quick growing and can provide the conditions required for some species, Sweet chestnut may provide a suitable alternative to oak. Horse chestnut may have good sap runs.*

- Create and maintain glades and rides.
- Ensure that there is continuity of linear landscape features such as lines of veteran trees and hedgerows. Bats and some invertebrates use such features as flight paths. (A gap of as little as 10 m in a line of trees can be enough to dissuade some bat species from flying along it as they travel between their roosts and their feeding areas.)
- Encourage flowers as nectar sources for invertebrates, eg hawthorn, composites, umbellifers and flowering ivy. A healthy invertebrate population will also support a healthy bat and bird population.
- Create and/or maintain associated habitats such as ponds and wetlands.

Figure 36. *An 'ideal' veteran tree site for wildlife.*

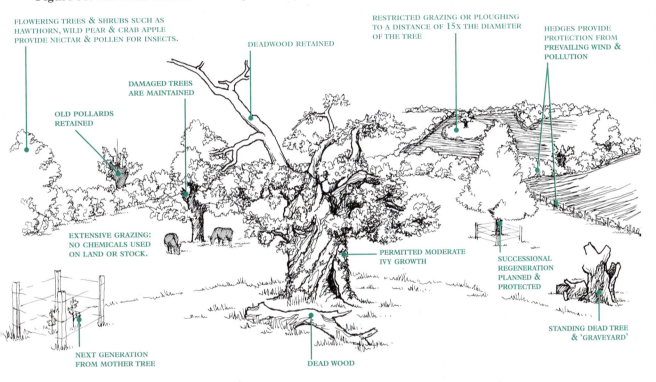

FLOWERING TREES & SHRUBS SUCH AS HAWTHORN, WILD PEAR & CRAB APPLE PROVIDE NECTAR & POLLEN FOR INSECTS.

DEADWOOD RETAINED

RESTRICTED GRAZING OR PLOUGHING TO A DISTANCE OF 15X THE DIAMETER OF THE TREE

HEDGES PROVIDE PROTECTION FROM PREVAILING WIND & POLLUTION

DAMAGED TREES ARE MAINTAINED

OLD POLLARDS RETAINED

EXTENSIVE GRAZING: NO CHEMICALS USED ON LAND OR STOCK.

PERMITTED MODERATE IVY GROWTH

SUCCESSIONAL REGENERATION PLANNED & PROTECTED

STANDING DEAD TREE & 'GRAVEYARD'

NEXT GENERATION FROM MOTHER TREE

DEAD WOOD

7.2.3 Creating cavities and decay in younger trees

If there is a lack of holes, crevices and decay it may be desirable to initiate some. This can be done in a variety of ways that are all best tried on younger trees, rather than veterans. Different groups of organisms have different requirements:

- Bats prefer deep narrow crevices.
- Birds mostly prefer holes rather than crevices, a variety of sizes will suit a range of species.
- Invertebrates use an almost infinite variety of decay, holes, crevices, etc. Consider drilling holes of various sizes into trees as well as making larger holes with saws or breaking off branches.

There is considerable scope for creating holes and initiating decay in trees (Figures 37 & 38). Customised 'boxes' can be made or more general cavities; experiment with what you have available.

Figure 37. *Suggestions for the design of artificial cavities.*

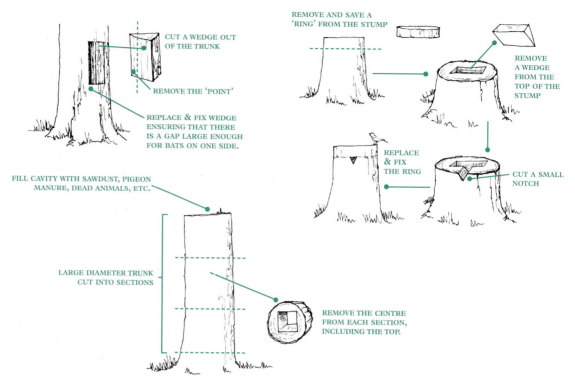

CUT A WEDGE OUT OF THE TRUNK

REMOVE THE 'POINT'

REPLACE & FIX WEDGE ENSURING THAT THERE IS A GAP LARGE ENOUGH FOR BATS ON ONE SIDE.

REMOVE AND SAVE A 'RING' FROM THE STUMP

REMOVE A WEDGE FROM THE TOP OF THE STUMP

REPLACE & FIX THE RING

CUT A SMALL NOTCH

FILL CAVITY WITH SAWDUST, PIGEON MANURE, DEAD ANIMALS, ETC.

LARGE DIAMETER TRUNK CUT INTO SECTIONS

REMOVE THE CENTRE FROM EACH SECTION, INCLUDING THE TOP.

Figure 38. *See colour plate page 90.*

Bat and bird boxes can be made or purchased. They should only be necessary if hollow trees are exceptionally sparce or specific boxes are needed for particular species, eg spotted fly-catcher or tawny owl. Nest boxes should be viewed as a 'stop gap' until suitable, natural, places are available. Take care when putting up boxes that you do not put them on part of a tree next to that used by rare invertebrates; the bats or birds may make a meal out of the threatened species.

In addition to the general requirements listed above, most groups have some very specific needs, which are dealt with below.

7.3 Epiphytes

7.3.1 Introduction

Various species of epiphytic plants are associated with old trees. The groups of greatest interest in this habitat are the mosses, liverworts and lichens but some vascular plants, such as ferns, can be epiphytic too. Among the bryophytes there is considerable regional variation in the numbers of rare and total species found. Epiphytic mosses and lichens can mostly be found throughout the year and are relatively easy to record and monitor although they require specialist identification skills.

7.3.2 Characteristics of moss growth on veteran trees

- In more highly polluted areas epiphytic mosses are found more on old trees than on young trees. They occur mostly low down on the trees and in sheltered positions (where they may be more susceptible to agricultural contamination).
- Veteran trees (and pollards) provide a variety of microhabitats, eg rain tracks, crevices, bark sheltered by protruding parts of the tree and exposed roots, where different species are able to survive.

> *The underside of horizontal branches and leaning trees keeps the plants dry and provides shelter from acid rain. Horizontal branches also have higher nitrogen inputs (eg as bird droppings) on the top and are good for certain species. Rot holes are good too when the rainwater persists and leaves accumulate. Mosses at the lip of the hole act as a wick so that it empties very slowly creating a rain track down the trunk for several days. A cavity with a small hole lower down may allow water to run out slowly for several weeks. Trees with such features are rare and support rare bryophytes.*

- Oak, ash, field maple and beech are the most important old trees for mosses, followed by hornbeam, sycamore, and hawthorn. Elm is also very good but few veteran elms survive except in the Scottish Highlands and Islands.
- Grazed wood-pasture allows light to the boles of the trees but the humidity is still relatively high. Dense woodland is often not so good for mosses despite the increased humidity that is liked by some species.
- Older trees have also had more time than younger trees for bryophyte growth to accumulate.

7.3.3 The characteristics of lichen growth on veteran trees

Two distinct lichen floras are associated with veteran trees. These are: old growth woodland species and those thriving in more open, drier, parkland type situations. Lichens are more sensitive to undergrazing and less to overgrazing than most groups of organisms.

7.3.3.1 Lichens and pollution

Lichens are very susceptible to sulphur dioxide and nitrous oxides and the location of the tree in relation to major sources of these substances is very important. Trees close to pollution sources tend to be species poor. Those in deep valleys are often better as the air passes over the top without penetrating (except when local pollution is trapped in a temperature inversion forming 'valley smog'). The tops of hills tend to be more polluted. The reduced levels of sulphur dioxide in southern England has enabled some of the more mobile species of lichen to recolonise but others, especially those typical of veteran trees, have very poor dispersal mechanisms.

7.3.3.2 The lichens of old growth woodland and veteran trees

- Rich woodland lichen floras depend on old growth woodland (ie stands older than 200 years) with veteran trees.
- Communities rich in important species take many years to colonise.
- The ideal conditions for woodland lichens are those with adequate light and shelter from drying winds.
- Ideal sites are those with a mosaic of dense and open areas; different species of lichen have different tolerances to exposure and light levels.
- Woodland lichens are adapted to low nitrogen levels so high nitrogen (ammonia) pollution from intensive farming is a serious threat.
- Slow growing species are found on virtually all types of bark, including acid bark but base rich bark is usually the richest habitat. Exotic tree species are rarely as valuable as natives, especially acid barked trees.
- Exposed heartwood areas may have special and rare floras.
- Ancient coppices are rarely good for lichens as they are more like young growth woodlands.
- In areas with a few old growth woods, local rarities can occur on occasional old trees within young growth woodland.
- Large populations of veteran trees are required for rich lichen floras to develop, as many species have very narrow and rare niches even in near-natural woods.

- A distance of as little as 2.5 km can prevent many species recolonising.
- Recovery time from clear-felling for most lichen communities is 200 - 300 years if there is nearby old growth. Communities of dry craggy oak bark take over 400 years.

7.3.3.3 Lichens of wayside and parkland veteran trees

- These communities are best developed on full lit trees with moderate enrichment from dung or dust. Well developed communities are absent from extensively grazed wood-pasture as the grazing levels are not high enough. Hence these communities are largely associated with human activity in Britain.
- The rarer species include slow growing ones and southern species at the edge of their range. Post mature and veteran trees are the richest.
- Unimproved grassland is the ideal habitat surrounding the veteran trees for these types of lichens.
- Base rich and mesic bark are the main habitats of interest but, unlike woodland lichen communities, acid bark is rarely of any interest. Exotic base rich barked trees such as Norway maple (a good elm substitute), walnut and tulip tree can be important. Sycamore can be rich but is not nearly as good as Norway maple. A few local specialists can occasionally be found on conifers.
- Landscape parks are now a major resource for this type of lichen flora but it was once common in agricultural areas with frequent old trees. The lichen communities have declined at least as much as those of the unimproved neutral grassland.

7.3.4 Management recommendations

The ideal management for lichens depends on the type of community present. Old growth woodland communities require light grazing and a mosaic of habitat structure. Parkland communities need more open conditions and can benefit from some exotic tree species being present.

7.3.4.1 Mosses and liverworts

- Don't drain or divert existing streams or damp hollows away from old trees as they may contribute to the overall humidity of the area (unless failing drainage threatens the tree).
- Do not block past drainage ditches in an attempt to increase humidity; if long established water levels are raised this can kill veteran trees.

7.3.4.2 Woodland lichens (Figure 39)

- Maintenance of old growth woodland containing veteran trees is crucial. Management of woodlands for commercial forestry is not very compatible with the conservation of woodland lichens. Traditional management such as grazing and pollarding can be compatible.
- Light grazing is beneficial in preventing uniformly shaded conditions. Deer alone may achieve this but the reintroduction of other grazers may be necessary. Where grazing has either ceased or declined dense shrub layers of, for example, holly or rhododendron can cause problems. Holly may be pollarded as it supports rare lichen species when not too shaded. This has been done in the New Forest as it also creates enough light to benefit lichens on old trees nearby. In addition, it perpetuates historic management on the site.
- Ivy, and other evergreens such as holly, can cause problems. Especially in ungrazed woods they can smother epiphytic growth. Ivy in the early stages of colonisation should be prevented from establishing on trees with high epiphytic value but old plants should be left. Ivy is beneficial to other forms of wildlife as it is a useful nectar source and provides cover. However, consideration might be given to its removal from trees in sites of high value for epiphytic plants, but low value for other groups of wildlife. Browsing of ivy on the lower trunks of the trees is the best and most natural method of ivy control.

- Be careful of suddenly opening up around relic lichen floras as lichens are very sensitive to light and humidity levels. Thus, clear small areas at a time and work on small groups of veteran trees at any one time to minimise local climate change. Also, the subsequent growth of the undergrowth (eg bramble) and tree seedlings in these areas may shade out lichens on trunks unless grazing levels are sufficiently high.
- Do not fell large, post mature exotic trees without checking for rare lichen species first.
- Some lichen rich communities are found on rocks (especially in upland areas) so these should be left if on site. (Other habitats can also support rich lichen communities e.g. the park pale or fence.)

Figure 39. *Site characteristics and management for old growth lichens.*

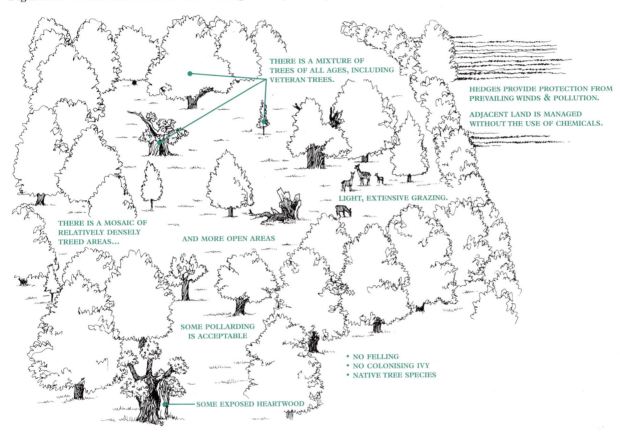

7.3.4.3 Wayside and parkland lichens (Figure 40)

- The maintenance of low intensity farming is the most important factor in conserving these lichen floras. In relict sites that are intensively farmed, ideally low intensity farming on permanent pasture should be restored.
- The application of farmyard manure or pelleted fertiliser is acceptable as long as it is not plastered on to the trees.
- Exotic trees with a base rich bark can be important and are much more acceptable in this habitat than in native woodland. Norway maple and walnut are especially good. New tree plantings should include all existing species of value on a site.
- When planting new trees, plant close enough to the existing veteran to maximise colonisation but not so close as to shade them out. As an example, plant three or four trees of the same species, or the same potential lichen flora as the veteran, in a semi-circle round the existing old one, avoiding the south side. A good guideline distance is 15 m.
- Parkland lichen floras are often able to respond relatively quickly if individual plants, showing signs of regression due to reduced light, are opened up sympathetically by trimming the canopy of the veteran tree or cutting surrounding trees.

Figure 40. *Site characteristics and management for parkland lichens.*

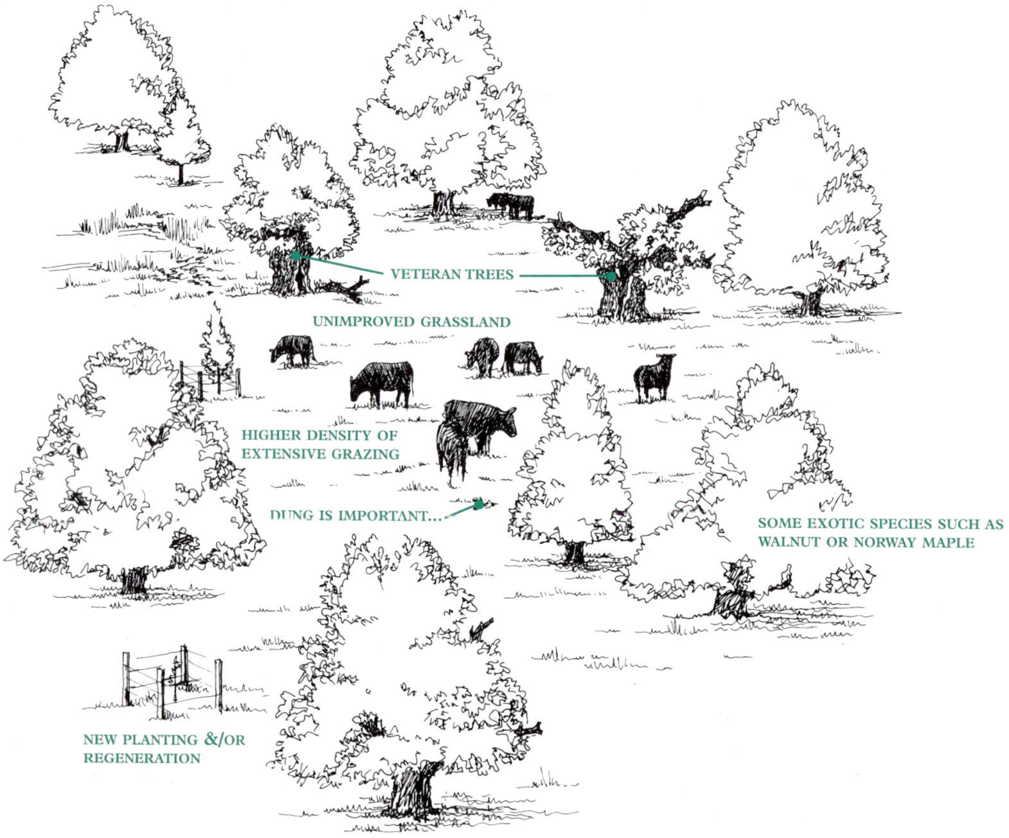

VETERAN TREES

UNIMPROVED GRASSLAND

HIGHER DENSITY OF
EXTENSIVE GRAZING

DUNG IS IMPORTANT...

SOME EXOTIC SPECIES SUCH AS
WALNUT OR NORWAY MAPLE

NEW PLANTING &/OR
REGENERATION

> *Translocations should be seen as a last resort where lichen trees have to be removed or are threatened. It can also be considered for boosting relict populations in damaged sites. Translocation has a low success rate and does not work for crust forming species. Try to find a tree as similar as possible to the host tree and use isopon glass fibre resin as glue ('bostic', 'evostick' and 'araldite' all might work but have been shown to induce some necrosis on the lichen). Try to take a part of the bark with the lichen and apply the glue to the bark. Recent experiments with lichen thallus lobes or small thallus fragments fastened to a new tree by nylon mesh staple to the bark have proved more promising than older techniques.*

Further reading: Adams (1996), Gilbert (1984, 1991), Harding and Rose (1986), Hodgetts (1989), Reed (1996), Rose (1976, 1991, 1993), Sanderson (1996b), Scheidegger *et al.* (1995).

7.4 Ground flora

While the main conservation value of sites with old trees is usually the trees themselves and the organisms living on them, the ground flora between the trees can be of interest in its own right. This aspect is also covered in chapter 5.

Examples of vegetation types include:

- Woodland - where the canopy of ancient trees (and associated younger trees) is dense enough that the ground flora is of a woodland type.
- Grassland - where the trees are spaced widely enough and the land is grazed so that pasture results.
- Heathland - as above but on heathy sites where heather dominated communities arise.
- Hedgerows - where ancient trees occur as part of ancient hedges.

On some sites with veteran trees a decline in grazing has caused a change in the ground flora and one option for future management is to reinstate the grazing. The merits of opening up woodland to recreate former wood-pasture must be carefully considered. Ideal management of the trees should take first priority, then the ground flora. Is the woodland flora of higher value than the pasture/heathland that will result? A specific rare or unusual species of plant that is not associated with the veteran trees can usually be protected/managed for without compromising the trees themselves.

Veteran trees in hedges are particularly vulnerable to damage by hedge trimmers, and flails. It is also necessary to ensure that some young trees are left in hedges to grow on and become ancient hedgerow trees of the future. Those at field corners often have a better chance of survival than those along the length of hedges.

Further reading: Harding & Rose (1986), Sanderson (1996a), Sanderson (in prep.)

7.5 Fungi

7.5.1 Fungi and veteran trees

Our understanding of the role of fungi in woodlands and especially woodlands of veteran trees has changed completely in recent years. Even as recently as the 1970s some people considered that the fungi rotting and hollowing out a veteran tree were detrimental to it and that fungal fruiting bodies on a trunk meant that the tree was soon to die. Fungi are fundamental to the growth of the tree and fulfil an important role at all stages in its life (Figure 41) see section 3.5 for more information about the role of fungi in the decay process. Remember that the fruiting body is just a tiny part of the organism; the rest, the mycelium, is very extensive but less often seen.

Fungi are beneficial to veteran trees and the organisms associated with them in the following ways:

*   **As mycorrhizal species. There are two different types**
    *   Ectomycorrhizae where the fungus sheaths the roots of the tree, eg fly agaric, boletes and amethyst deceiver.
    *   Endomycorrhizae or vesicular arbuscular (VA) where the fungus penetrates the cells of the roots. These are much smaller fungi and less easy to see.
    Mycorrhizae enhance the ability of the tree to take up phosphorus and nitrogen and in return the fungi gain carbohydrates. The fungi probably also help the tree withstand drought, pollutants and pathogens. It has been shown that trees grown without their mycorrhizae are very poor specimens. Most plants have mycorrhizal fungi associated with them, not just trees.
*   **As nutrient cyclers. Fungi are one of the main agents of decomposition for both leaves and woody material on the woodland floor and inside the tree. They recycle nutrients from the dead plant material and make them available again. The products of the fungal decomposition are taken up by invertebrates, plants, and even the trees themselves when aerial roots from the tree grow into the rotting trunk.**
*   **As a consequence of the decay inside the tree, especially in veterans, the conditions are created for a range of animals and plants including saproxylic invertebrates, birds and bats.**
*   **Through an interaction between the fungi and the trees, zones of decay within living trees become 'compartmentalised'. This creates a range of different microhabitats within a single tree. The presence of one species of fungus may inhibit the growth of others and may confer protection from more aggressive species.**
*   **The fruiting bodies of the fungi provide a food resource for invertebrates and mammals.**

Not all fungi are beneficial to trees. Some are pathogenic and can kill trees, but they are in the minority. Pathogenic fungi tend to be far less common in natural woodlands than in plantations and isolated trees in ornamental situations. There are a few important points to be remembered when a pathogenic species is suspected:

- Many pathogens take hold when the tree is already stressed for some reason, eg by drought or shading.
- Some pathogenic species are in fact species groups in which individual species vary considerably in their behaviour, eg of the honey fungi, *Armillaria mellea* can be pathogenic but *Armillaria gallica*, which looks almost identical, is more frequently found in woodland situations and is only weakly pathogenic.
- Some species produce copious fruiting bodies on freshly dead wood but are not responsible for the death of the tree, eg oyster fungus.
- Heart rotters are not usually pathogens; they are just causing the decay of the dead wood in the centre of the tree, eg *Laetiporus*.
- Pathogenic species are more likely to be a problem when trees are in monospecific stands or more or less isolated within agricultural land or gardens, rather than in natural woodland situations.

However, there are some situations when fungi can at least contribute to the decline of veteran trees. An important example is the growth of *Bjerkandera adusta* (a sap wood colonising species) on pollarded hornbeam in Essex. This highlights the need to carry out work on a few trees at a time and alter the programme in the light of experience. *Bjerkandera* seems to be encouraged by the drying out of exposed wood following complete crown removal.

Like other organisms some fungal species are rare and threatened. Many of those found hollowing out old trees have restricted distributions as do some of those found on undisturbed grassland areas surrounding old trees. Of the 447 macrofungi on the British Red Data Book list all but 50 are from ancient woodland and lowland wood pasture. Most of them are wood decomposing or mycorrhizal.

In view of their importance, and in some instances conservation status, measures to conserve fungi should be considered.

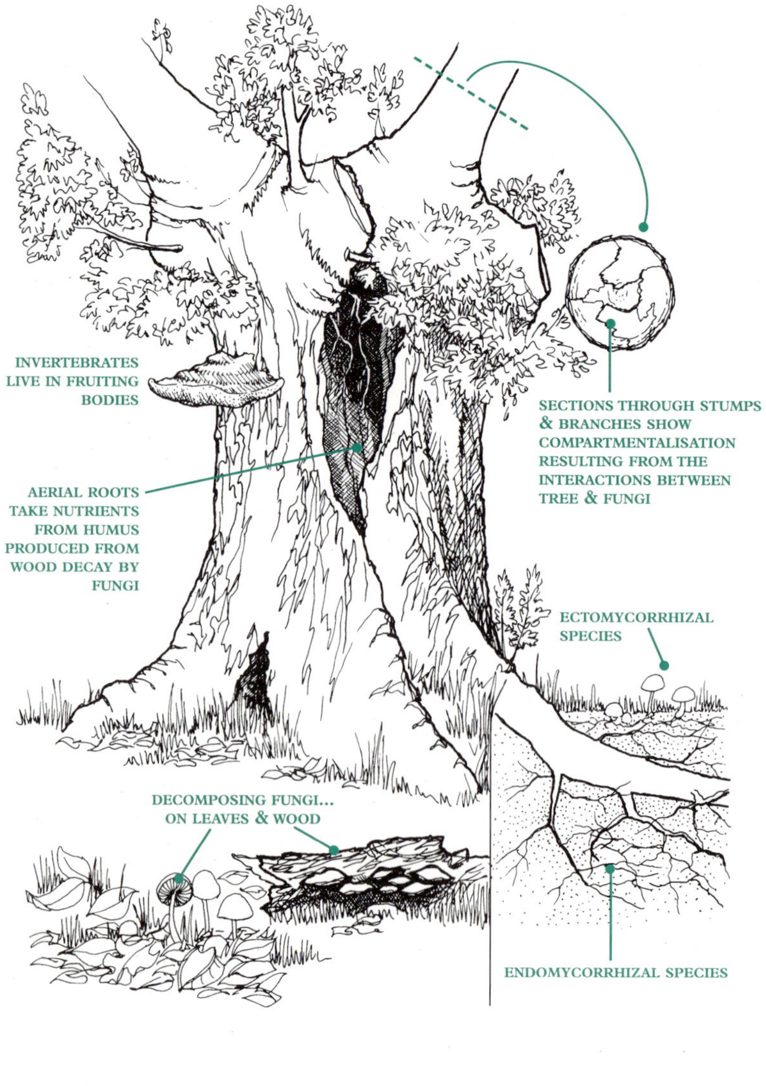

INVERTEBRATES LIVE IN FRUITING BODIES

AERIAL ROOTS TAKE NUTRIENTS FROM HUMUS PRODUCED FROM WOOD DECAY BY FUNGI

SECTIONS THROUGH STUMPS & BRANCHES SHOW COMPARTMENTALISATION RESULTING FROM THE INTERACTIONS BETWEEN TREE & FUNGI

ECTOMYCORRHIZAL SPECIES

DECOMPOSING FUNGI... ON LEAVES & WOOD

ENDOMYCORRHIZAL SPECIES

Figure 41. *The importance of fungi for veteran trees.*

Figure 1. *A veteran oak tree in Lincolnshire.*

Figure 2 *Veteran hawthorn pollards at Croft Castle (Hereford).*

Figure 4. *Wrest Park, Bedfordshire. Early 17th Century sketch by Peter Tilleman showing mature trees incorporated within the immature formal landscape. Courtesy of Bedfordshire and Luton Archive.*

Figure 5. *A veteran ash pollard in the Lake District.*

Figure 6. *A lapsed pollard at Ashtead Common (Surrey).*

Figure 7. *Ancient wood-pasture with veteran trees at Farmcote (Cotswolds).*

Figure 9. *A veteran tree killed by car parking around it. Calke Abbey (Derbyshire).*

Figure 13. *Sweet chestnut trees at Croft Castle (Herefordshire). A standing dead tree is in the foreground and stag-headed trees behind.*

Figure 14. *The cut surface of a log showing internal boundaries in the timber.*

Figure 15. *Chicken of the woods* Laetiporus sulphureus, *on an oak tree.*

Figure 21. *An ash pollard at Goswold Hall (Suffolk) last cut about 20 years ago by the owner (in photograph) using the saw he is holding.*

Figure 22. *A veteran hornbeam pollard at Hatfield Forest (Essex) being cut for the second time in recent years.*

Veteran Trees: A guide to good management ▮ Page 85

Figure 23. *Old growth woodland in the New Forest (Hampshire), grazed extensively.*

Figure 24. *Whittlewood Forest (Northamptonshire), a deer park with improved grassland.*

Figure 24a. *Veteran trees killed by a bracken fire at Ashtead Common (Surrey).*

Figure 25. *Ploughing too close to a veteran tree. The tree has also had its lower branches removed, which is also detrimental.*

Figure 26. *Gradually opening up around a veteran oak surrounded by conifers, Ethy (Cornwall).*

Figure 28. *The base of a large beech tree, fallen apart and left to decay naturally, Windsor Forest (Berkshire).*

Figure 30. *A landscape with veteran trees, Watenlath, Lake District.*

VETERAN TREES
I N I T I A T I V E

Figure 31. *'Coronet' cuts on a veteran oak tree, Ashtead Common (Surrey).*

Figure 32. *Veteran tree on an ancient monument, at Hailes Abbey.*

Figure 33. *A veteran sycamore pollard outside the Fox and Hounds at Foss Cross, Cotswolds.*

Figure 38. *An example of an artificial cavity, Windsor Forest (Berkshire). Note the birch log that has been stood up against the living tree to create more standing dead wood.*

Figure 43. *Wasp mimic cranefly* Ctenophora flaveolata *is a Red Data Book species, dependent on soft, decaying heartwood of very large veteran beech trees.*

Figure 44a. *A treespade being used to move a young oak tree at Ashtead Common (Surrey),*

Figure 45. *A large tree guard in use, Hatfield Forest (Essex).*

Figure 46. *Young, newly created pollards at Hatfield Forest (Essex).*

Figure 47. *An avenue of veteran sweet chestnut trees at Croft Castle (Herefordshire).*

Figure 48. *The Major oak at Sherwood Forest (Nottinghamshire). The tree has been fenced to reduce the trampling effect of visitors, and also mulched.*

Figure 50. *An ash tree at Hatfield Forest (Essex) after the first cut, a second cut being planned.*

Figure 51. *Restoration pollarding a veteran beech at Burnham Beeches (Buckinghamshire). Cutting work has been completed on this tree.*

Figure 52. *New holly pollards in the New Forest (Hampshire).*

Figure 53. *Recently cut hornbeam pollards at Knebworth (Hertfordshire).*

Figure 54. *A fallen veteran willow in Lincolnshire that is regrowing well.*

VETERAN TREES
I N I T I A T I V E

7.5.2 The management of veteran trees for fungi

There are various ways in which veteran trees and the land around them can be managed to favour fungi, including those species beneficial to the survival of the trees.

- Avoid over collecting/damage to fruiting bodies. There has been considerable discussion and conflict over this issue in recent years.

Collection of fungi

While picking of fruit bodies may not be harmful to the fungal organism itself, it is detrimental in other ways. First it may reduce the chance of sexual reproduction and long range dispersal of spores. Secondly it deprives many animals of a food source. While for mammals fungi may just be a supplementary part of the diet, some invertebrates rely on specific species of fungi and live only in the fruiting body. In sites where recreational pressure is high, damage to fungi may be considerable. Occasional picking is not likely to be a problem, but intense and repeated collection of certain species can be detrimental. On sites with good populations of old trees, removal of fungal fruiting bodies, especially from the old trees, should be avoided. A code of conduct has been produced for collecting fungi, giving useful guidelines (English Nature 1998).

- Excessive trampling by people or grazing animals is detrimental to mycorrhizal species.
- Fertilising of old trees should not be carried out. Inorganic fertilisers and lime have a detrimental effect on mycorrhizal fungi. Trees in fertilised areas may appear healthy for many years but in times of stress succumb more easily.
- Light grazing (eg by cattle, ponies or sheep) in woodland may encourage increased numbers of species and numbers of fruiting bodies of mycorrhizal fungi.
- Removing leaf litter from around amenity trees may encourage root disease causing fungal species. This is because their natural competitors are suppressed.
- Try to avoid causing urban trees any undue stress, which may upset the natural balance of fungi species and mycorrhiza, eg by trenching or laying tarmac over the roots.
- If an aggressive *Armillaria* species is found, trees nearby can be protected by constructing a barrier in the soil, using a phenolic soil drench, or winching out or grinding infected stumps and roots and then burning them. This is a sensible procedure only in urban situations.
- Excessive ivy cover on trees can smother the formation of bracket fungi (see also section 7.3).

Further reading: Alexander, Green & Key (1996), English Nature (1998), Green (1991), Ing (1996), Lonsdale (1999a & b), Marren (1992), Rayner (1996).

7.6 Invertebrates

7.6.1 Introduction

The numbers of species of saproxylic insects (ie those dependent on dead and decaying wood) are far higher than those of other groups of organisms associated with old trees and a remarkable proportion of these are of rare and uncommon species. However, on many sites the invertebrate fauna is very poorly known in comparison to birds etc. Most groups of invertebrates have species associated with old trees but those with the largest numbers of species are the beetles and flies (figure 43). Some are active decomposers of wood, assisting with nutrient recycling, others feed on fungi and there are also predators and parasites that are specialists in the dead wood habitat. Throughout Europe, saproxylic species have been identified as the most threatened community of invertebrates.

Invertebrates have very special features making them different from other organisms.

- Their cycle is often annual; thus making populations very vulnerable to poor years and lack of continuity of habitat. (However, some saproxylic species can have long larval stages due, for example, to the poor nutritional quality of the food. Stag beetles take up to five years to reach the adult stage.)
- Different stages in the life cycle of the same species may have very different requirements and habits. Saproxylic species are often either predatory or feed on nectar or pollen as adults, usually in very different places from where the larvae develop.
- Some are extremely specialised. For some, there are likely to be very few trees with the precise conditions required, even on sites with a good veteran tree population.

Stag beetles

The law prohibits the sale, and advertising for sale, of stag beetles. Guidance on what to do if larvae are found in a decaying stump is given in a leaflet Stags in Stumps *available from the People's Trust for Endangered Species or the English Nature enquiry service.*

A good tree for invertebrates (Figure 42):

- Dead wood in the crown - hot dry wood supports a limited but specialised range of species.
- Decay columns - brown rot and soft white rot are especially valuable.
- Rot holes in a variety of sizes, dampnesses and stages of decay, eg some water-filled and others dry and containing tree humus.
- Partly decomposed wood, burrows and cavities, resulting form actions of other saproxylic species.
- Sap runs or fluxes, where the sap oozes out of the tree.
- Fungal fruiting bodies and fungi present under the bark etc.
- Damage to the bark, eg lightning strike.
- Broken branch stubs that are good for invertebrate access, eg for egg laying.
- Nectar source nearby.
- Fallen branches left to lie near the tree in partial shade.
- Living tissue (ie the tree is alive) so that it can continue to produce more dead wood and shade the dead wood already on the tree.

Figure 42. *A good tree for invertebrates.*

1 **Major Deadwood**
Sunbaked, aerial deadwood, desiccated wood (longhorn beetles).

2 **Upper Crown Limb - Small Cavities**
Dry rot holes - birds, bats roost (hornets nest).

3 **Crown Limb - Large Cavity**
'Brown' rot (stiletto flies, cardinal click beetle, darkling beetles, barn owl roosts).

4 **Fungal Growth on Limb**
Fungi on bark (wood awl flies, false ladybirds).

5 **Snag/Stub**
Large surface area for egg laying and fungi (cardinal beetle).

6 **Bark with Fungal Infection**
Fungi on bark (cardinal beetles, wood awl flies, false lady birds).

7 **Suspended Broken Limb**
Shattered end provides large surface area for egg laying and fungi.

8 **Weak Fork with Included Bark**
Nest (birds, squirrels, rove beetles, micromoths).

9 **Water-Filled Rot Hole**
Water-filled rot hole (hoverflies, water beetles).

10 **Flux on Bark**
Established sap run (sap beetles, hoverflies and fungus gnats).

11 **Scar Tissue from Old Wound**
Damaged loose bark (bark beetle, false scorpions and spiders).

12 **Bracket Fungi**
Heart rot prepares wood for invertebrates; (fungus gnats, shining fungus beetles).

13 **Delamination of Wood**
Fungi/invertebrates (cardinal beetle, sap beetle).

14 **Subsiding Major Limb**
May lead to shattered stub habitat.

15 **Fallen Limb**
Fallen timber habitat: leave in partial shade.

16 **Lightning Strike**
Burnt wood (flat bugs, false weevil, smoke flies).

17 **Fungal Colonisation of Root**
Damaged loose bark: (bark beetles, false scorpions and spiders).

18 **Basal Cavity**
Hollowing trunk (cardinal beetles, lesser stag beetle, crane flies).

19 **Rot hole in Trunk**
Soft rot (lesser stag beetle, rhinoceros beetle, crane flies)

20 **Root Damage from Browsing**
Soft rot (stag beetle, hoverflies).

Figure 43. *See colour plate page 91.*

7.6.2 Good management for invertebrates

- Do nothing to damage those features illustrated in the diagram as being good for invertebrates (including draining or damaging decay holes).
- Keep as much dead wood as possible on site, preferably all of it.
- Leave any wood on site that shows any sign of internal decay or loose bark. (The most valuable dead wood for invertebrates is that which is decaying from the inside out rather than from the outside in (eg from rot holes and heart wood).)
- Very small diameter timber - brash, twigs and small branches - are more useful if accumulated into piles in the shade of the tree canopy or of bracken or brambles.
- If timber has, for some reason, to be removed from the site or destroyed then do not mature it on site. Remove it as soon as possible after cutting and certainly before the end of April. If timber (due to be removed eventually) must remain on site after this time it should be covered. This will speed up the seasoning process but more importantly it will prevent colonisation by invertebrates that will then be removed and destroyed.
- Be aware that some dead wood provides hibernation sites for some species and these will be affected by winter removal of wood. Some species continue to develop right through the winter period too.
- Allow some undergrowth, eg brambles or bracken, to scramble over and protect dead wood from desiccation, especially in grazed areas, but not so much that it is detrimental to the trees (eg causing a fire hazard or competing for water).
- Do not carry out management work on all the trees at the same time. Ensure that there is always a range of dry and, in particular, humid and shady conditions in the same areas. This is true of street trees and those beside rivers as well as in woodlands or parkland.
- Ensure that there are adequate nectar sources in open sunny conditions. Spring flowering shrubs are important and open structured flowers, eg hawthorn and umbellifers, are best as the insects do not need specialised mouthparts to feed from them. Be careful as some cultivated varieties (especially double flowered varieties such as red hawthorn) may either flower at different times of the year or have no nectar. Ivy is also useful as it provides nectar in the autumn and also provides a deadwood habitat in its own right, it is not parasitic and poses no threat to the trees.
- It is particularly important to retain the same tree species composition on the site when planting younger trees and encouraging the regeneration. This is because different species of tree have different invertebrate faunas associated with them.
- The translocation of dead wood invertebrates from one site to another should be considered very carefully before action is taken. Often the detailed ecological requirements are not known well enough to be sure of success. Advice needs to be sought from specialist entomologists before translocations are attempted. Note that they also confuse the true status of the species and its distribution. Guidance on establishing species on new sites is available from the Joint Committee for the Conservation of Invertebrates.

7.6.2.1 A good site for invertebrates

Sites that are especially good for saproxylic invertebrates tend to have:
- A large number of old trees especially of native species. (Oak is especially important for beetles, and beech for flies.)
- Plenty of dead wood on the ground.
- A long historic continuity of dead wood and old trees (this may be established by researching historical documents).

- Trees showing signs of decay, rot holes, sap runs and dead wood in the crown, ie those features often associated with old trees. The dead wood that develops from rot holes and heart wood is especially valuable. See also comment on the location of dead wood given above.
- Trees that are native broadleaves (except in Scotland and areas such as the Breckland where Scots pine is also important). However, very old specimens of exotic species, such as sweet and horse chestnut and very large sycamores can also provide valuable habitats. Such old specimens should not be removed on conservation grounds simply because they are not native.
- A good mixture of structure - for example open grassy areas, deep woodland.
- Good nectar sources, eg trees, bushes and herbs with open accessible flowers.

If a site does not show all of these features it is still worth bearing in mind the requirements of invertebrates even in an individual, isolated tree.

7.6.2.2 Insect Surveys

Although surveys are the best way of finding out what is living in a particular site it is more or less impossible to carry out a comprehensive survey. Nevertheless it is important that survey work is done on sites to assess their conservation value. Evaluation techniques (mainly using beetles) have been developed see Harding & Rose (1986), Fowles et al. (1999).

One of the problems of sampling saproxylic faunas is that often the habitat is destroyed or damaged whilst searching. There is an increasing number of methods available where this can be avoided (eg searching nectar sources in season or using traps such as Owen emergence, Malaise or flight interception traps). In many instances, however, a skilled entomologist searching by hand is still the most valuable way of finding important species.

Detailed surveys can be expensive to commission but some invertebrate societies and groups can be persuaded to visit sites with potential, and valuable information and contacts can be built up in this way. See the English Nature Species Handbook for further information.

A code for the entomological investigation of dead wood has been drawn up to aid managers and owners of sites and also to give guidelines for entomologists. See English Nature (1994), Fry & Lonsdale (1991) and Key & Ball (1993).

Further reading: Alexander, Green & Key (1996), English Nature (1994), Fowles (1997), Fowles, Alexander & Key (in press), Fry & Lonsdale (1991), Green (1995a), Hammond & Harding (1991), Harding & Alexander (1993), Harding & Rose (1986), Key (1993, 1996), Key & Ball (1993), Kirby (1992), Marren (1990), McLean & Speight (1993), Speight (1989), Watkins (1990).

7.7 Birds

Wooded country in general is very important for birds, with more breeding species found than any other major habitat in Britain. The complex structure of woodland is important for birds as is the abundance of food. Many species also like the more open aspect of a parkland type habitat. Thirty five per cent of British woodland bird species require holes or crevices to nest in and this is precisely the habitat provided by ancient trees. A few birds can excavate their own holes but most rely on ready-made ones. Ivy on trees is also valuable for birds.

The main bird nesting season is between March and July and ideally no work should be done on veteran trees during this period. This time is also best avoided from the point of view of the tree but if bats are present it may be desirable to do tree surgery in March. In this instance survey work may be necessary to ensure that there are no birds nesting in the tree.

Further reading: Fuller (1995), Smart & Andrews (1985).

7.8 Reptiles and amphibians

Reptiles and amphibians make use of cavities and loose bark on veteran trees as resting places. Veteran trees are, for example, very valuable for grass snakes. General advice on management for reptiles and amphibians can be found in English Nature (1994).

7.9 Mammals - Bats

7.9.1 Introduction

All of the 16 British species of bat depend on trees to some extent although the degree of dependence varies according to the species. Some, (including the brown long-eared and the pipistrelles) have been able to adapt to roosting in houses at certain times of the year, but others such as noctule and Bechstein's bat are strongly associated with tree roosts.

7.9.2 Ways in which bats use veteran trees

Veteran trees provide important habitat for bats throughout the year (Figure 44) as follows:

- summer roosts - species such as barbastelle, Daubenton's and Natterer's bats need holes and crevices in trees, for roosting in and giving birth to their young in the summer months.
- winter roosts - pipistrelles and brown long-eared bats tend to use trees more in the autumn and winter. Spaces beneath loose bark or ivy may be used, as well as holes and crevices.
- year round roosting sites - noctule, Leisler's and Bechstein's bats all need tree holes both for breeding in the summer and hibernation in winter.
- source of food - All British bats feed exclusively on insects, and because areas of ancient woodland and parkland are rich in insects they are particularly important to bats as foraging habitats. Even species that do not roost in trees, such as the serotine, will nevertheless forage around them.

Signs of bat occupation of trees

A bats' roost in an old woodpecker hole or a crevice may exhibit one or more of the following signs of bat occupation:

- Dark staining around the entrance from the oil in the bats' fur.
- A streak of dark staining running down the trunk, where the droppings have been washed out of the hole.
- Droppings (which crumble into a fine powder when rubbed) stuck to the trunk or on the ground below the roost.
- Scratch marks from the bats' claws around the edges of the hole.
- Bats heard squeaking in the middle of the day during warm weather.
- Flies and other insects buzzing round the entrance, attracted to the droppings inside the roost.

However, there may be no sign at all from the outside that bats are present in a tree. This is particularly true in winter, when the bats may be hibernating deep within the trunk of a hollow tree. Even in summer, telltale external signs of bat occupation are often absent. Bats can use very narrow crevices and the smaller species can fit into a crack of less than 15 mm. If in doubt, get a tree checked by an experienced bat worker.

Figure 44. *A bat's year.*

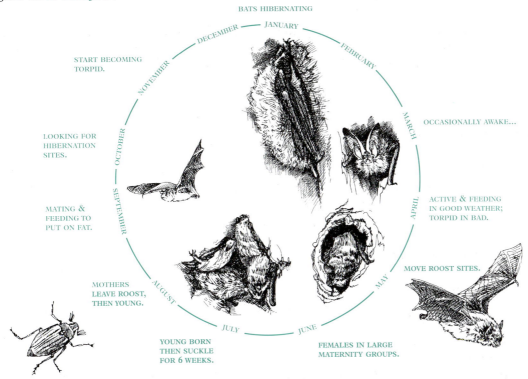

BATS HIBERNATING

DECEMBER — JANUARY

FEBRUARY

START BECOMING TORPID.

NOVEMBER

MARCH

OCCASIONALLY AWAKE...

LOOKING FOR HIBERNATION SITES.

OCTOBER

APRIL

ACTIVE & FEEDING IN GOOD WEATHER; TORPID IN BAD.

SEPTEMBER

MATING & FEEDING TO PUT ON FAT.

MOVE ROOST SITES.

MAY

AUGUST

MOTHERS LEAVE ROOST, THEN YOUNG.

JULY — JUNE

YOUNG BORN THEN SUCKLE FOR 6 WEEKS.

FEMALES IN LARGE MATERNITY GROUPS.

7.9.3 Recommended procedure for working on veteran trees that may contain bats

- If a tree is a known bat roost it is mandatory to seek advice from the relevant statutory nature conservation organisation (English Nature, Countryside Council for Wales, Scottish Natural Heritage or Environment & Heritage Service (Northern Ireland) before doing any work.

Note. It is an offence to intentionally destroy a known bat roost whether or not the bats are present at the time. A roost is defined as "any structure or place which is used for shelter or protection" this includes trees used by roosting bats.

- If the tree is not a known roost, but contains holes or crevices, has loose bark, or a thick covering of ivy, it should ideally be surveyed for bats by an experienced bat worker prior to any work. Contact the relevant statutory nature conservation organisation or your local bat group to request a survey.
- Sometimes, even experienced bat workers have difficulty in deciding whether a tree contains bats - particularly in situations where access to some parts of the tree for close inspection may be restricted. If the presence of bats is possible but not proven, the following precautions should be taken:
  - Try to restrict work to the periods mid-March to May, and September to November. This is because during the summer young may be present and unable to fly, and in the winter hibernating bats will be slow to arouse and unable to escape. Also, disturbance of bats in winter depletes their fat reserves, and can reduce their chances of surviving to the following spring. Note that this coincides *exactly* with the times not recommended for cutting trees in chapter 4. This can be overcome by cutting the trees just before their leaves emerge in early spring if bats are suspected. The March - May period also clashes with peak bird nesting time.
  - Be careful that any cracks held open by the weight of a branch do not contain bats. Such cracks may close up when the branch is removed and squash the bats.
  - If a branch has holes and crevices and cannot be retained it should be lowered to the ground gently, not dropped.

7.9.4 If bats are discovered during the course of work:

- You must stop all work at the first safe opportunity and seek advice from the appropriate statutory nature conservation organisation - even if the bats have managed to fly away. The statutory nature conservation organisation will normally arrange for a member of your local bat group to come out to attend to any grounded bats, but it is a good idea to know the number of your local bat group in case of emergency. If you are unable to find it, the Bat Conservation Trust should be able to give it to you.

- If any of the bats are injured or torpid, gently place them in a canvas bag or a lidded box while awaiting help. Be careful as bats are very delicate. Use soft leather gloves to handle them; bats do not normally bite but may do so if injured and in pain. Make a note of exactly where the bats were found; this is important if they are to be released after veterinary treatment.

- If bats are present inside wood felled from the tree, but thought to be unharmed, try to enclose them in their roost by covering the entrance until the bat worker arrives. It may be possible for the piece of wood containing the roost to be lodged in the crown of a nearby tree, for the bats to leave of their own accord, but the bat group will advise on this when they arrive. (They will also examine the bats to confirm that they are uninjured as it can often take considerable experience to tell.)

Further reading: Bat Conservation Trust (1997), Holmes (1996, 1997, 1998), Hopkins (1998), Mitchell-Jones & McLeish (1999).

7.10 Other mammals

Other mammals also benefit from holes in veteran trees, including holes at ground level. These include the introduced grey squirrel, which can be harmful to hole-nesting birds such as hawfinch and even tawny owl. Other native species using tree holes include red squirrel, common dormice and mustelids such as polecat, pine marten, weasel and stoat. None of these species are specific to veteran trees.

7.11 Time of year to work on the trees

The ideal time of year to do work on veteran trees, according to the organisms associated with them, is illustrated below:

	Jan	Feb	Mar	Apr	May	Jun	Jul	Aug	Sep	Oct	Nov	Dec
Trees*	✔	✔	✔	X	X	X	✔	✔	X	X	✔	✔
Bats	X	X	✔	✔	✔	X	X	X	✔	✔	✔	X
Birds	✔	✔	X	X	X	X	✔	✔	✔	✔	✔	✔
Epiphytes	✔	✔	✔	✔	✔	✔	✔	✔	✔	✔	✔	✔
Fungi	✔	✔	✔	✔	✔	✔	✔	✔	✔	✔	✔	✔
Invertebrates	✔	✔	✔	✔	✔	✔	✔	✔	✔	✔	✔	✔

✔ - Suitable time
X - Unsuitable time

*The time of the year that is best for the veteran tree is extremely important. If the tree dies then it no longer contributes continued habitat for the other groups.

VETERAN TREES
I N I T I A T I V E

| Chapter 8 | The next generation of Veteran Trees |

8.1 Introduction

One of the biggest threats to our rich heritage of veteran trees is the absence of a next generation to replace them when eventually they die. Linked to this is the lack of suitable conditions for wood decay for the wildlife dependent on these conditions. Saplings may replace the veterans in many years' time but a crucial need is for trees of middle age, that are nearly veterans. These in particular need protecting. If there are no 'near veterans' there is a need to create suitable decay conditions on younger trees. It is also very important that the life of existing veterans is extended to enable a large overlap in life spans.

8.2 Lack of the next generation of trees

A lack of new/young trees is normally caused either by grazing pressure (including deer and rabbits) eliminating any natural regeneration or by regeneration not occurring owing to shading, eg by planted conifers, dense bracken or rhododendron.

When considering options for establishing the next generation of trees bear in mind that naturally grown trees tend to survive better than transplanted trees because of the way that the roots develop. Thus natural regeneration should be the first option if possible. The use of tree shelters often increases the chance of survival of the young trees but may suppress lateral branching, which is important for open grown trees if they are to have high wildlife value. The methods for ensuring the next generation (in order of preference) are:

1. *Natural regeneration from existing trees (very old trees may not produce as much viable seed as young ones).*
2. *Seed from the site scattered on the ground and grown on* in situ.
3. *Seedlings grown, in a nursery, from existing veteran trees and planted out.*
4. *Seedlings grown, in a nursery, from trees elsewhere on the site and planted out.*
5. *Seedlings from other local sites.*
6. *Saplings moved from elsewhere on the site.*
7. *Saplings from elsewhere, ie not from the same site.*

Ideally the next generation should be:

- of similar genetic origin to that already on site, ie of local provenance. This can only be assured if you collect the seed yourself from the trees with the characteristics you want to perpetuate or if natural regeneration occurs. (When collecting seed it is best to do so from existing veterans as these may have the genetic predisposition to live for a long time).
- of similar species composition to that of the old trees, unless there is a need identified for some faster growing trees, eg birch (or sweet chestnut) to provide, for example, saproxylic conditions sooner;
- planted/encouraged before the ancient ones are lost. (An oak tree grown from seed may take 200 years before it starts producing dead wood.) This also means that the lives of the veterans must be extended as long as possible;
- used to extend the existing site boundary if possible;
- planted/encouraged continuously so that trees develop at different stages and provide a range of age classes. A single cohort of young trees will repeat today's problems in 300 years' time. However, trees do not need to be planted every year. Gaps of 10 years or so between cohorts is fine;
- planted/encouraged in ways that take the landscape character into account;
- planted in groups, eg three (if possible), to allow for some losses.
- not planted so close to veterans that the young trees grow up to interfere with the older ones. Open grown trees at maturity may have a canopy spread of over 30 m.

Encouraging natural regeneration in grazed areas

- Alter the grazing regime to reduce the density or remove the animals either from the whole area or, from small patches that can be rotated if necessary. In this instance grazing pressure may need to be relaxed for several years in order for the seedlings to reach a height where the animals cannot reach the tops. The land inside the fence may still need to be managed. Consider fences that allow some grazers (eg sheep) through but not others (eg horses). Without any control the growth of vegetation surrounding the young trees can swamp them.
- In a large area allow rough patches to develop, which the animals are unable to get to and allows seedlings to 'get away', eg bramble patches round fallen trees. Be careful that the exclusion of grazing is not detrimental in woods that have been grazed, as this can allow dense growth of species such as holly that shade out everything else. Do not clear away rough patches when they form.

8.2.1 Solving the lack of the next generation

8.2.1.1 Planting/translocating trees

Growing seed from the site (or elsewhere) for planting is very time consuming. However, one way round this is to involve the local community, especially children, who can also help with the long-term care of the trees. Children planting seeds in pots enhances the chance that they will be looked after and this is the focus of the Trees of Time and Place project. Some trees, notably willow, grow well from cuttings and this can be a valuable way of ensuring subsequent generations. Large cuttings of up to 3 - 3.5 m cut from coppice or pollard willow usually grow well.

Translocation of trees at Ashtead Common

In common with many sites, there is a lack of suitable maidens to pollard within the relict wood-pasture areas on Ashtead Common. In the scrub grassland the regeneration of oak is much better but the younger trees often have to be removed as they are growing above a gas pipe or under high voltage cables. These young oaks are moved into the wood-pasture areas by means of a treespade (Figure 44a). The trees are lifted out of the ground together with a large root ball that is then wrapped in hessian and chicken wire and transported to the planting location. Trees are moved at an age of between 10 and 15 years of age so that they will be above the bracken and less susceptible to deer damage. Rabbit guards round the trunks protect them from deer fraying. Survival rate has been about 85%, the deaths probably being caused by very dry summers and attempts to move trees that were too large.

Figure 44a. *See colour plate page 91.*

When planting out seedlings/saplings their aftercare needs to be considered too:

- If sites are grazed the trees must be protected with guards appropriate to type of animals. The use of chemical deterrents on leading shoots to restrict browsing is another possible option.
- Big tree guards (Figure 45) allow the tree to develop a more natural shape while still giving protection.

Figure 45. *See colour plate page 91.*

- Tree shelters may help growth in the early stages but they may need to be removed/replaced and do not mean that trees can just be forgotten.
- Trees may need to be cared for, eg removing competing weeds and other saplings, watering in drought situations.
- Ideally they need to be planted in forest soil, to encourage associations with mycorrhizal fungi.

For full details on how to plant trees see British Trust for Conservation Volunteers (1980).

8.2.1.2 Where to plant/translocate young trees

In woodland, the position of the new trees is usually not especially important. However, in wooded commons or in the general countryside you should consider how the planting will affect the existing regional character with regard to species, position and landscape character. While it is beneficial to increase the area of a site this should not be done at the expense of other, valuable habitat, types eg heathland or unimproved grassland, or landscape features. Even within existing sites with veteran trees it is important to make sure that conditions such as soil type are suitable for the tree species you are about to plant, and not to plant up good habitats.

In a more formal landscape, especially one that has been designed with specific objectives in mind, it is vital that any planting is done in sympathy with the historical design. In these sites you should get advice from a historian or landscape architect. Points to bear in mind are:-

- Avenues should not have their shape broken up.
- Views that were an important part of designed landscapes, eg from the main house, from high points or features of interest on the estate, should not be interrupted by new plantings. (Remember that a few years after planting the trees will be taller!)
- Look carefully at the layout of trees on the estate. Are they in formal clumps, groves, regular blocks, belts or groups of a specific number? Try to emulate this pattern if possible. To the flora and fauna how the trees are arranged is generally of less importance than the fact that they are there.
- Try to copy the existing species composition on the site. Ideally, native species should be in the majority although non-native species may be integral to the design.

8.3 Lack of abundant wood decay

8.3.1 Lack of imminent veterans

Encouraging natural regeneration and the planting of saplings may help to fill the generation gap on some sites, but a more immediate problem is ensuring that trees that are nearly veterans survive to become veterans. There is an urgent need to ensure that trees in their prime are retained into old age. Many of the characteristics normally associated with veteran trees (for example, dead wood in the canopy, decay holes, loose bark and sap runs) are found in younger trees and these should be cherished, not seen as detrimental.

Even when a new young generation of trees is growing on well, there may be a considerable gap before these young trees gain the characteristics of veterans. Continuity of these features is very important for the wildlife value of the trees. This situation can be remedied in several ways but most require some form of active management. Leaving dead wood on the ground will help, but for sites with a good saproxylic fauna it is not a replacement for dead wood and rot holes in standing trees.

Some practical solutions to this problem exist:

- Leave older trees out of the felling rotation cycle - this solution is only possible when plantation trees of suitable species have been grown round the old ones (see section 5.3.2 for details).
- Cutting pollards from maiden trees - this will ensure older trees in the future and may also create dead wood in the bollings of the young pollards, especially if trees are cut older than the age at which pollarding would usually have commenced, (see below for more details).
- Damaged trees or those in need of reducing should be made into pollards, or reduced in a similar way, rather than being felled.
- Plant or encourage faster growing species in addition to replacing the longer lived species. These can, to some extent, provide suitable conditions, but should not be viewed as a complete replacement. For example, birch harbours a good range of species, sweet chestnut is a good bridge of oak communities, horse chestnut has a good range of species and good sap runs and sycamore supports a reasonable range of species).
- 'Damaging' young or middle aged trees will encourage the development of decay.

8.3.2 Methods for creating veteran tree characteristics on younger trees

A number of methods can be used to encourage veteran characteristics on younger trees. This type of work should be done on young or mature trees not veterans. Methods that retain a live tree are better than those that kill the tree as live trees will continue to provide conditions for saproxylic species. Dead trees are a short-term measure. These methods include:

- cutting the tops off trees that are likely to respond to pollarding. It is best to do this when the intention is to create new pollards and then to accept that some will not survive long as a source of dead wood. Cutting large trees will help create more decay communities than young ones;
- ring barking. This kills many trees slowly thus allowing rot to develop. This is preferable only when it is desired to get rid of trees, eg undesirable species or those too densely planted, as killing trees does not produce long-term continuity of habitat. It is best done in woodland as trees in the open 'season' rather than decay;
- making holes in live standing trees to initiate rot;
- inoculating heart wood rotting fungi into healthy trees;
- putting the tree under stress;
- using explosives to produce shattered ends to branches and 'natural' looking damage. (This option should only be carried out by experienced specialists with a licence;)
- using herbicides to kill standing trees;
- constructing special saproxylic 'nest boxes' see section 7.2.3;
- deliberately damaging the bark to induce decay earlier or simulate sap runs (squirrel damage can have the same effect);
- breaking branches, rather than sawing them off flush, or creating 'coronet' ends;
- increasing the water retention in forks and crowns of trees by drilling holes.

8.3.3 Creating new pollards

Creating new pollards (Figure 46) can provide continuity of the dead wood resource and potential ancient trees for the future. It can also provide continuity of historical or landscape interest.

The relative importance of these points will vary according to the site but you should consider:

- **The species of tree to cut.** Try to perpetuate the existing range of species already on the site, although there may be good reasons for pollarding some other species, eg sweet chestnut, see sections 7.2.2 and 7.6.2.

- **Height of cut.** Try to maintain roughly the same height and shape of tree as the existing pollards, but be aware of the management implications of this decision.
- **Where to pollard.** The site may be divided into areas of different historical management, eg by wood banks, park pales. Wherever possible maintain the distribution of new pollards much the same as the old ones to keep in sympathy with the landscape. There may be a need to create new pollards in different areas if there are no suitable young trees close to the ancient pollards.

8.3.3.1 How to create new pollards

- Cut as young as possible, when the tree has attained the desired height.
- In most species of tree, the creation of a new pollard is easy. The stem of the sapling is cut at the desired height. Depending on the species (see appendix 4) branches may or may not need to be left below the point of cutting. Bear in mind that growth will come from below the cut so cut a little higher than you want.
- Even in shade-tolerant species, like beech, adequate light must reach the stem.
- Trees up to 40 cm in diameter can usually be cut without leaving a leader except for beech and oak.
- For more difficult species, eg beech, some branches must be retained, preferably two or more to provide a degree of balance; cut according to the shape of the tree.
- Prolific growth may follow (especially in the case of lime, willow, etc) which needs to be pruned back if the area is not grazed, to ensure that growth is encouraged from the top.
- Once the tree has been cut, try to ensure that it is maintained, eg by cutting in the future. A 10 - 20 year cycle is probably appropriate for most situations.
- For most species the older the tree the more reluctant it is to grow following pollarding. Prepare for this by leaving branches on (leave more branches on the greater the diameter of the tree, and more for beech relative to oak and ash).
- In the USA urban pollards are cut so that one branch is left on at the junction of the cut and the main stem. This is removed after 1 - 2 years (Coder 1996).
- If, for ash in particular, shoots do not appear in the first spring after cutting, don't panic! They may come later.
- Remember that new pollards that die are sources of dead wood. If you have scope for making plenty, don't worry if some die, keep them as wildlife habitats.
- Remember that what is successful on one site may not be on another. Variations in soil type and rainfall make a difference. Always err on the side of caution.
- For time of year to cut, see guidelines for cutting old trees.
- Cutting height should be determined, at least in part, with the grazing animals in mind. The browse line is approximately 1.3 m for fallow deer, 2.1 m for cattle, 2.7 - 3.0 m for horses. Cut at least 30 cm above the eventual browse line.
- Bigger, older trees can be cut like pollards to help close up the generation gap (eg 100 - 150 years old) but lower branches must be retained in this situation. Most often (oak and beech) the subsequent growth will be from the retained branches rather than being new ones but it has a similar effect in prolonging the life of the tree and creating suitable conditions for decay to occur.
- On sites with public access it is desirable to inform the public what you are doing and why.

Figure 46. *See colour plate page 92.*

8.3.3.2 Subsequent cuts

If an appropriate cutting regime is known for a site it would seem sensible to follow this. If it is not known, it seems likely that most cuts were made at 10 - 15 year intervals. Willow was probably cut more frequently. In street trees cuts can be made much more frequently and the timing depends on the degree of growth and the situation of the tree.

Many 'natural' pollards were created during the storms of 1987 and 1990 but most will not be maintained as pollards in the future. Natural pollards can also be caused by squirrel damage (eg on young beech trees).

Further reading: Alexander, Green & Key (1996), Atkinson (1996), Barwick (1996), Battel (1996), British Trust for Conservation Volunteers (1980), Coleman (1996), Edlin (1971), Forbes & Warnock (1996), Kerr (1992), Key & Ball (1993), P. Kirby (1992), Mitchell (1989), Rackham (1986), Sanderson (1991), Sisitka (1991a, 1991b), Speight (1989), Tubbs (1986), Watkins (1990).

| Chapter 9 | Dealing with conflicting management priorities |

9.1 Introduction

In a situation where there are conflicting ideas on how to manage a specific tree or a site containing veteran trees:

- Ensure the needs of veteran trees are built into site management plans.
- Gather as much accurate and up to date information as you can about the situation. If necessary carry out survey work/historical research.
- Meet those with an interest in the particular issue and don't rely on assumptions about their likely views.
- Approach the situation with an open mind and be honest about the relative merits of the site/tree. Encourage others to be too.
- Weigh up the relative importance of the site/tree for the various interests, assign relative weights if this helps to analyse the situation.
- If an easy solution cannot be found, look at the possibility of a compromise that does not result in significant loss of interest for conflicting issues.
- Visit other sites with similar conflicts and learn from them.

More information about specific veteran trees or sites with veteran trees is available in the following data base (for contact addresses see appendix 6):

- The Register of Parks and Gardens of Special Historic Interest in England - Prepared by English Heritage. Published as a set 46 county volumes and in most reference libraries and local planning authorities. Also available for purchase from English Heritage. An updated edition is being prepared and it is hoped will be available in digitised form from 2001.
- Champion trees (Tree Register of the British Isles).
- Invertebrate Site Register (Contact the relevant Statutory Nature Conservation Organisation).
- English Nature is collating a database giving pointers to information sources on the wildlife and heritage information of packs and wood-pastures.

9.2 Specific potential problems in veteran tree management and suggestions for how to overcome them

9.2.1 Habitat or hazard

Problem: A veteran tree that is decaying and has several dead branches in the crown can be a concern to safety.

Discussion: It is inevitable that old trees will have decayed wood, cavities and dead branches, all features which enhance their habitat value. Contrary to much popular opinion, this does not mean that they are necessarily dangerous. All trees have the potential to cause damage to differing degrees. A risk assessment needs to be made for each situation together with the type and likelihood of damage . There is, therefore, no such thing as a perfectly safe tree and some degree of risk will remain even in a sound tree with no defects. The task is to evaluate the hazard that the defects pose and the risk of damage to people or property if the tree, or part of it, fails, and to take appropriate action to reduce the risk to an acceptable level. Health and safety legislation recognises that it is unreasonable, and in many cases not feasible, to eliminate risk. It is possible to have habitat and hazard. It is not a question of either or. Owners and managers may be alarmed by reports of individual judgements in cases involving trees, but the law does not require a choice to be made between trees, their habitat and people and property. It is a matter of assessing whether the dangers posed by the tree could have been anticipated, and whether these dangers could have been countered by means of moderate and reasonable remedies. Those who fail to understand this, pose a great threat to the veteran trees of Britain.

9.2.2 Nature conservation and designed landscapes

Problem: A designed landscape that has matured and is now in decline has increased in nature conservation value. Priorities for management and repair could affect either the historic or nature conservation value.

Discussion: In many cases this conflict is perceived through a misunderstanding or lack of appreciation of the other parties' objectives. For example, perpetuating the pattern and vistas in the landscape may be a priority for one party, the species selection and management of trees for nature conservation may be a priority for another party. Here, the presumption of retaining veteran trees, their survival and ensuring that there is a new generation of trees is desirable for both parties.

Where a genuine conflict of interests occurs it is important that each party states their ideal objectives. The situation should be approached with an open mind and the will to succeed. Issues should be put in perspective by understanding and appreciating the value of the site/tree in relation to others, for the various interests. Where necessary a compromise must be reached between the loss of the historic fabric and loss of habitat.

9.2.3 Conflicts of interest between the needs of different organisms

Problem: A site may be important for rare lichens preferring an open canopy and rare invertebrates needing a shady environment.

Discussion: An accurate evaluation is needed of the relative importance of these groups and where they are found on the site. Management can usually accommodate both or, it may be found that the conservation status of one species is considerably higher than that of the other. It is very unlikely that a single tree would be the home for two extremely rare organisms requiring opposing management.

9.2.4 Commercial aspects of the site conflict with the ideal management for nature conservation reasons

Problem: In commercial forestry or agriculture veteran trees may take up land which could be more productive.

Discussion: The retention of individuals and groups of overmature and veteran trees, as well as dying and dead trees, is recommended in the UK Forestry Standard (Forestry Authority, 1998) as well as the Forest Nature Conservation Guidelines (Forestry Commission 1990). It is also recommended to identify younger trees to become the veterans for the future. Loss of revenue may not be as great as expected and being informed about the conservation value for the old trees may be enough to ensure their survival. There is no evidence that retained dead wood in broadleaved forestry plantations puts commercial crops at risk (Winter 1993). Grants may be available to manage veteran trees in a variety of different situations, (Woodland Grant Scheme, Countryside Stewardship, etc).

9.2.5 Increasing the productivity of the land conflicts with ideal tree management

Problem: There is a desire to increase the fertility and productivity of grazing land, in ways that are detrimental to old trees. This may be by applying fertiliser, chemical sprays or ploughing close to the trees.

Discussion: Grants may help with this (eg Countryside Stewardship), or look at alternative farming systems (eg organic) which may attract set-up grants and a premium on products.

VETERAN TREES

9.2.6 Public access causes detrimental effects to the trees

Problem: Pressures to increase public access may increase the need for safety work.
Discussion: Draw up a clear safety policy for the site which states the importance of the trees and the methods used for surveying and implementing work needed. Consider methods of visitor management such as zoning, re-routing paths, re-locating car parks or picnic sites, or changing ground vegetation (ie long grass, dead hedges) to encourage people away from high risk areas. Talk to organisations that have experience of solving similar problems, eg The National Trust.

9.2.7 Aesthetically appealing or ugly and untidy

Problem: Trees that appear wonderful, interesting and beautiful to some people are grotesque and ugly to others. While the sentiments of owners and managers have undoubtedly caused the demise of ancient trees in the past (and also saved many too) this should not normally be a cause of conflict today. The retention of dead wood on the ground is still sometimes removed because it is viewed as being 'untidy'. This is an especially important issue in historic parkland where public access has a significant effect on management.
Discussion: Education and information is often the key here. Pointing out the age of the tree and what it has 'seen' is usually a better starting point than the number of insects and fungi it houses (see also section 6.3).

Veteran pollards in Epping Forest

In the period leading up to the Epping Forest Act in 1878 negative attitudes had grown towards the pollarded trees. They were seen as symbolic of a particular way of life and indicative of past mismanagement and over-exploitation of the Forest. One result of the Act was to change the emphasis of Forest management from protection of Commoners rights, to the provision of a recreation area. Although the Act protected the pollards many influential people (including members of the Essex Field Club) desired a 'natural' appearance and saw no place for pollards. A journalist described the hornbeam pollards as 'short, shabby, scrubby, indescribably mean and ugly'. Even the president of the Field Club, a biologist of some renown, thought it "desirable that many of the pollards should be removed' and saw 'no reason why in time they should not all be replaced by spear-trees." From Dagley & Burman (1996).

9.2.8 Exotic or native species

Problem: There is sometimes a desire to plant exotic species on a site (eg for timber purposes or in a designed landscape or garden). Any potential impact of this will depend largely on the species concerned.
Discussion: Weigh up the likely response of the species and the naturalness of the site. What is appropriate for an ancient semi-natural woodland may well be different from that for a formal garden. Introducing *Rhododendron ponticum* should be opposed (it can contribute to the death of veteran trees by competing for water when growing around them and also prevent any regeneration owing to its heavy shade), but specimen trees in a formal garden setting are unlikely to present any problems. Even on sites with no particular historic interest some exotics can provide a useful 'stop-gap'. Fast growing species such as sweet chestnut (Figure 47), and even sycamore, may provide suitable conditions for saproxylic species if there are no suitable aged native trees on the site. In designed landscapes some exotic species may need to be planted to provide historical continuity. If possible, plant or encourage native species grown from local stock.

Figure 47. *See colour plate page 92.*

9.2.9 Financial constraints restrict the amount of work that is desired

Problem: The ideal management is too costly to achieve.

Discussion: Do the best you can and prioritise the management so that important work is done first. Remember that a long-term view is necessary when dealing with trees. Not all the work will need doing at once so a 20-year plan may be quite good enough and a short period of time in terms of the life of the tree.

VETERAN TREES
I N I T I A T I V E

Chapter 10	Public access and Veteran Trees

10.1 The benefits and disadvantages of public access

The trend towards greater public access in woodland can bring considerable benefits in terms of greater understanding of woodland processes as well as giving enjoyment to the visitors themselves. However, in some situations the health of veteran trees can be threatened because of this. Probably the greatest single threat to veteran trees in Britain today is their felling on the grounds that they are a hazard to public safety. This complicated aspect is covered in a separate leaflet produced by the Veteran Trees Initiative. There are other ways in which veteran trees may be jeopardised by people. It is ironic that many of these threats arise because the public is appreciating the trees, perhaps for their historic value or unusual appearance.

- Public access can cause damage to trees and associated organisms in the following ways:
- Intentionally by:
    - burning through intended fires or accidental ones;
    - vandalism, for example damaging trees, damaging the bark, graffiti.

- Unintentionally by:
    - climbing on trees, which damages the bark so the tree is unable to transport food and water efficiently;
    - trampling round trees, which can in extreme cases compact the soil affecting water uptake and mycorrhizal fungi (Figure 48);
    - collecting of specimens by the public or survey work by naturalists;
    - removal of dead wood;
    - car parking under old trees (causing compaction);
    - marking trees with waymarks for trails, etc.

Figure 48. *See colour plate page 92.*

10.2 Solutions

Some solutions to these problems are:

- Inform the public about the value of trees through signs, leaflets, guide books, guided walks and nature trails.
- Fence the tree to reduce compaction, etc. This is not a recommended solution but may be necessary for key feature trees.
- Draw up a collecting code for the site or adopt one already in existence.
- Try selling a limited amount of woodland products (ideally those of lower conservation value) and use the money for site management. Visitors then feel that they are contributing. BUT ensure that only the surplus is sold.
- Relocate car parks to open areas and/or use shrubs for shade and screening or trees not intended as future veterans. Encourage visitors to walk to interesting areas not drive as close as possible.
- Discourage access right up to trees where there is a specific problem, eg by using dead hedging (especially prickly plants).
- Erect raised walkways around key feature trees.

Chapter 11   Legal aspects of management for/of Veteran Trees

11.1 Introduction

There is a range of legal obligations on those owning, managing or working on ancient trees. It is your responsibility to establish which of these applies in your situation.

1. *Tree Preservation Orders*
2. *Conservation Areas (towns and villages)*
3. *Felling licences*
4. *SSSI/NNR/SAC*
5. *Scheduled Ancient Monument*
6. *Wildlife and Countryside Act - Bats*
7. *Wildlife and Countryside Act - other species*
8. *Hedgerow legislation*
9. *Owners/occupier liability*
10. *Health and safety (operational work)*

11.2 Tree Preservation Orders

These are placed on trees by the local planning authority for amenity reasons and usually apply to individual trees (occasionally groups of trees or areas). Permission is required from the local planning authority for any work to be done on the tree (including pollarding and crown thinning) and heavy fines are given for not gaining permission. Veteran trees can, however, still be felled if they are considered unsafe. It is possible for anyone to request that a TPO is put on any particular tree (contact your Local Authority) although it may not always be carried out. They are usually placed on trees of landscape interest but there are some exemptions.

11.3 Conservation Areas (towns and villages)

Any trees in a designated conservation area of a town or village are protected in the same way as trees with TPOs. If in any doubt, contact your Local Authority.

11.4 Felling licence

This is required from the Forestry Commission for felling more than $5\,\mathrm{m}^3$ in any calendar quarter, eg 1st January to 31st March. If less is felled then no more than $2\,\mathrm{m}^3$ can be sold. There are various exemptions and the Forestry Commission should be contacted for details. It may be necessary to clear surrounding woodland or commercial plantations from veteran trees and a felling licence would be necessary in these situations. Note that permission is not required for pollarding but is for cutting coppice when the stems have a diameter of more than 15 cm.

11.5 SSSI/NNR/SAC

For sites with a designated conservation status, work on old trees (both surgery and felling) needs to be approved by the relevant statutory nature conservation agency. The best way is usually to draw up a management plan, which is then approved. Then only work not included in the plan will need further approval.

11.6 Scheduled Ancient Monuments

If work affects a Scheduled Ancient Monument, or is in the vicinity of one, Scheduled Monument Consent may be necessary. Advice should be sought from the appropriate Inspector of Ancient Monuments for English Heritage, CADW or Historic Scotland.

11.7 Wildlife and Countryside Act - Bats

All bats and their roosts are protected under the Wildlife and Countryside Act (Schedule 5), 1981 (as amended) and are also included in Schedule 2 of the Conservation Regulations, 1994. The roost is protected even if the bats are not present at the time. If bats are found or a roost is suspected the relevant statutory nature conservation organisation should be contacted immediately. Information on bats is also available from the Bat Conservation Trust.

11.8 Wildlife and Countryside Act - other species

During the course of work on veteran trees other species may be encountered which are covered by legislation. It is an offence to take or destroy an egg laid by a wild bird and this includes destroying nests with eggs in during the course of tree surgery. The law with regard to birds is quite complex; some species have greater protection than this and others are not protected. If in doubt, check the Act (obtainable from HMSO).

Other species are protected too, for example badgers and their setts.

11.9 Hedgerows

Important hedges are protected under the Hedgerow Regulations (1997). Removal, including removal of trees, requires permission for certain categories of hedges. Your local authority should be contacted before any work is done.

11.10 Owner/occupier liability

All trees can be dangerous if they fall on people or property and there is a perception that old trees are more dangerous than young ones. The owner of a property has a duty of care to people coming onto his land (even if they are trespassing) and should take all reasonable action to make sure that his trees are safe. Ultimately, the only truly safe tree is one felled at ground level but this is not an option that should be followed with ancient trees unless there is really no other solution.

It is necessary for the owner therefore to look at the risk associated with his trees, ie the chance that if it fell it would cause damage. He should ensure that he has a system in place for assessing and surveying trees and for dealing promptly with any trees that are hazardous and in high-risk areas.

This complex issue is considered in more detail in a separate leaflet produced by the Veteran Trees Initiative.

11.11 Health and safety at work

The health and safety regulations for occupational workers and other persons in the vicinity are extensive. Work on ancient trees can be extremely dangerous. Ensure that those working on such trees are approved contractors (eg the Arboricultural Association has a list) who take safety issues seriously. These contractors are certificated in climbing, chainsaw use and using a chainsaw at height. If using 'in house' staff they should be properly trained (eg in the use of chainsaws and in climbing trees) and wearing appropriate personal protective equipment. Do not let volunteers use machinery or carry out work on old trees unless you know that they are fully trained and protected.

Chapter 12 Keeping records

12.1 Why record?

The number of veteran trees in England and Wales is still unknown although most of the major sites with many veteran trees are well known.

Tree surveys have been started for certain areas and on particular sites but it is very important to extend this. Information about tree populations and population dynamics is very limited at present. The production of a standard recording form and method of survey should improve the situation and it is hoped that future surveys will ensure that data collected is compatible with the Veteran Tree Initiative recording form. This system is equally suitable for single trees or large populations. A copy of the form for individual trees is enclosed with this publication, and can be photocopied. Forms for recording many trees can be obtained from English Nature. When more results are obtained it should start to be possible to look at regional differences and highlight potentially valuable sites. Repeat surveys are then needed to assess, and possibly model, mortality rates in key sites. Work of this type has been carried out at Duncombe Park.

As well as recording the condition of the trees at a point in time it is also necessary to record any work done and how the tree responded. It is essential that future generations of people are aware of what has been done in the past and the rationale behind it.

For the better management of the site it may be beneficial to have veteran trees mapped, especially on sites with many individuals. This can be done by compass and tape but increasingly now Geographical Positioning Systems linked to palm top computers are being used for this purpose and are able to provide more accurate positioning. These is scope here for computerised information of each tree (eg from the tree recording forms) to be stored on computer too. As with all computer systems, make sure you keep a back up copy and a paper version too.

Surveying populations of old trees can be time consuming but the information gathered is valuable today and will be especially valuable in years to come. Surveys already completed for sites can yield important information about the age structure, mortality rate and condition of trees.

Further reading: Clayden (1996), Forbes & Warnock (1996), Read, Frater & Noble 1996).

Information to record when working on veteran trees

Date.
Type of tree (ie pollard, lapsed pollard, veteran, maiden).
Species.
Approximate age of tree.
Type of tool used to do the cutting.
Method of cutting (ie slanted, rip, flush cut).
Length of stub left.
Type of bark on each stem and any other characteristics.
Number of branches removed/left.
Situation of the tree, especially the amount of light (exposed or sheltered, etc).
Response 1 year after cutting.
Response 5 years after cutting.

Take photographs before and after cutting and 1 and 5 years after.

12.2 Tree tagging

In areas with many ancient trees it may be necessary to mark or label individuals so that they can be individually identified. Knock-in timber tags are of limited value, rather better are 6 cm stainless steel tags and aluminium nails. If long nails are used, and hammered only 2 cm into the tree, the tags are able to swing freely and the tree can grow a considerable amount before the tag is engulfed in the bark. An alternative is to attach the label to 6 cm of stainless steel or plastic coated wire, and knock a 3 cm aluminium nail all the way into the tree. While the nail will quickly become surrounded by the bark the wire will not. Plastic tags can be used but become brittle and do not last well. Aluminium tags are easily damaged by squirrels, birds and people. Although other methods are available, nails are usually needed to attach tags safely to hard, rugose bark. Aluminium is less likely to prove toxic to the tree and is more sympathetic to chainsaws. Galvanised nails are suitable for the tree but are harmful to lichens.

Modern techniques such as computer chips and transponders are worth watching for in the future. None are routinely used on veteran trees and there are still problems such as securing them in the trees, locating them again and pre-selecting numbers (most use large numbers of random digits).

There is not yet a really good reliable and permanent method for tagging trees.

Further reading: Fay (1996), Fretwell & Green (1996), Key & Ball (1993).

Chapter 13 — Funding and advice

13.1 Introduction

It will have become apparent from previous sections that active management is often needed on sites with old trees. This may not necessarily mean work on the old trees themselves but often on the land surrounding them or to encourage a new generation of trees. This work can be costly.

Various organisations give grants for habitat work that includes ancient trees but further details should be obtained to check up to date facts and figures. Present sources include:

- Countryside Stewardship (funding for pollarding and tree surgery as part of a site, including farms).
- Environmentally Sensitive Area (ESA) payments.
- Forestry Commission Woodland Grant Scheme can include work to benefit veteran trees in certain wood-pasture situations.
- Heritage Lottery funding is a possibility for survey and historic landscape work.
- NNR section 35 (1) (c) - Available for National Nature Reserves owned or managed by organisations other than English Nature.

Another option is to make the trees themselves pay. This can be as woodland products such as firewood, charcoal, venison, or domestic animals but these should be by-products of the work, not a reason for doing it. Neither the veterans themselves nor the associated flora and fauna should be compromised as a result of the work. For example, dead wood should not be removed from the veterans to generate income. However, twisted and knotted wood that has no timber value can yield a high income from wood turners, who like burrs and spalted wood.

It is extremely difficult to estimate costs for pollarding and work on ancient trees. In some situations it may be possible to cut 10 trees in a day or only one, depending on the difficulty.

13.2 Sources of advice and getting work done

While awareness of the values and management problems associated with ancient trees is growing it is not safe to assume that all woodland advisors and tree surgeons are competent and have experience in this area. Even the Arboricultural Association approved contractors, although they may be very experienced, may not have ever worked on old trees let alone be able to give detailed and accurate advice. Appendix 6 gives details of some organisations that may be able to help with funding or advice.

It is recommended to use Arboricultural Association approved contractors, and the Ancient Tree Forum may be able to help locate people in your area experienced in veteran trees. Be wary of using the same contractors for advice and for doing the practical work, they may be rather over zealous about the amount of work that needs doing.

If in doubt about the advice you have been given try to find someone else who has dealt with a similar situation before.

Suitable questions to ask a contractor

1. *What are the constraints to working on a tree?*
    -   Legal constraints such as Tree Preservation Orders, Conservation Areas, or felling licences.
    -   Wildlife value.

2. *What signs would indicate that a tree has a high wildlife value?*
    Holes, cavities, water-filled cavities, loose bark, staining, bracket fungi, etc.

3. *What might be living in the tree?*
    Bats, birds, insects, etc.

4. *What would you do if you found bats or nesting birds during the course of the work?*
    Refer to chapter 7 for suitable replies.

5. *How can a tree be made safe, causing minimum damage to its wildlife value?*
    Remove the target, crown reduction, propping, cable bracing, etc.

The impression that you should get is that the contractor would not cut the tree unless absolutely necessary and that he would do the minimum of tree surgery necessary to achieve the required aim. The contractor should also have adequate knowledge of the wildlife value of veteran trees and know what to do if a protected species were found during work in progress.

Glossary

Figure 48a. *Features of a Veteran Tree.*

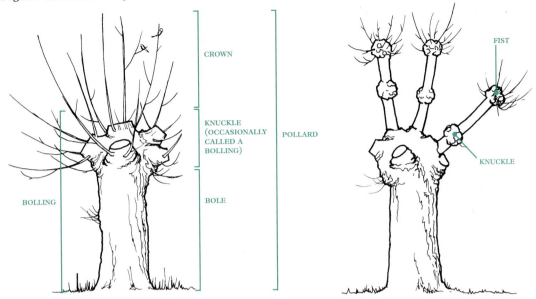

ancient tree -	The final stage in a tree's life.
adventitious buds -	Secondary buds arising in already existing tissue, often as a result of damage.
aerial roots -	Adventitious roots, developing from bark tissue on the above-ground parts of a tree, sometimes into a decaying trunk.
bolling - (see figure 48a)	The permanent trunk and stubs of a pollard consisting of the knuckle and the bole. It can take many different shapes depending on the form of the tree. Sometimes used for the knuckle alone.
bole - (see figure 48a)	The main trunk of a pollard.
branch bark ridge -	The area of raised bark tissue that forms at the junction of a branch and the main stem.
branch collar -	The swelling at the base of a branch formed when the growth of the branch is disproportionately slower than that of the main stem. The term is also used for the growth pattern of the cells of the main stem, around those of the branch, even if no swelling is visible.
brown rot -	That where the cellulose is degraded but the lignin is only modified. Also called red rot, orange rot, etc.
bundle -	A tree that has, naturally or by planting, originated from two or more seedlings of the same or different species, in close proximity. As the individuals grow they become closely pressed together.
burr -	A tumour-like swelling on a tree resulting from any number of causes, sometimes associated with epicormic growth.
buttress -	A swelling or spur at a base of a trunk where a root differentiates into the stem. Collectively the buttresses form a flare.
callus -	An undifferentiated mass of cells, for example on the surface of wounded living plant tissue. Also a fold of differentiated wood and bark that forms around a wound on a tree.
cambial zone -	A multiple layer of meristematic cells, which divides to form the increments (rings) of bark and/or woody tissues.
cambium -	The layer of cells that develops into the cambial zone (see above), existing as a single layer during dormancy. The vascular cambium forms wood on its inside and bark on the outside, whereas the cork cambium lies within the bark and forms corky cells on its outside and sometimes a secondary cortex on its inside.

canker -	An area where the bark and the cambium have been killed by disease; it is usually occluded by new bark and wood forming around its edge.
canopy -	The coverage of leaf area of one or more trees. **Open canopy** where the trees are spaced apart, **closed canopy** where they overlap. Also the uppermost layer of woodland.
case hardened -	When the surface of exposed wood dries rapidly and seals in internal moisture. The underlying wood may be functional or decaying.
cavity -	Hole in a tree caused by the removal or destruction of wood.
cellulose -	The main component of plant cell walls; a carbohydrate composed of long filaments, made up of glucose molecules.
chase -	Unenclosed land where wild animals were preserved for hunting, usually owned by a subject rather than the Crown.
coarse woody debris (CWD) -	An American expression referring to dead wood with a diameter of more than 2.5 cm.
compartmentalisation -	The physiological isolation of columns of wood within the tree.
coppard -	A tree coppiced and, later in life, pollarded.
coppice -	An area of trees cut near ground level and left to regenerate from the stool. Often cut as a block, usually as part of a fairly formal rotation. Also used as the verb to coppice and as an intransitive verb meaning the response of the tree.
coppice stool -	See stool
copse -	A wood used for cutting coppice (variant of coppice) but also used more loosely for a small wood.
cork cambium -	A layer of cells within the bark, laying down corky cells on the outside.
coronet cut -	Irregular cuts made in the stub left after a branch has been removed, the aim of which is to give the cut surface a more natural appearance.
crown - (see figure 48a)	The spreading branches and foliage of a tree.
crown thining -	The systematic pruning of small diameter branches throughout the crown. With the aim of reducing weight or wind load.
crown reduction -	The reduction of branch length in part or whole of the crown.
cubical rot -	A form of brown rot, where the decayed wood breaks apart in cubes.
cuboidal rot -	As above.
dbh -	Diameter at breast height. Usually 1.3 m above the ground unless the tree shows abnormal swellings at that point.
deadwood -	Wood which no longer fulfils any function for the tree. It may still be attached or have fallen from the tree.
dead wood -	Wood that no longer contains living cells. Includes deadwood (as defined above) and also heartwood etc., which may have a structural function in the tree.
decay -	The chemical breakdown of wood by micro-organisms.
dormant buds -	Those formed during the development of the current year's shoots but which do not develop unless later stimulated to grow.
doted -	See spalted.
dottard -	A tree that has lost its top or branches, is dead and in a state of decay.
dozed -	see spalted.
dieback -	The death of a part of the tree, usually from the periphery inwards.

dysfunctional wood –	Wood that has lost all or part of its original function (eg it can be structurally important but not conductive).
early-wood –	The wood produced by trees in their main flush of growth, in the spring. The wood may have better water conduction ability than late-wood but is often less resistant to dysfunction and decay.
endophytic fungi –	Fungi living within plant tissues without causing overt disease.
epicormic growth –	Literally, growth 'upon stem', initially appearing as twiggy growth apparently from the bark surface. There are two types, that from dormant buds and that developing adventitiously.
epiphyte –	A plant or lichen growing on a plant.
flush cut –	A cut that cuts into the branch bark ridge, which injures the trunk.
flux –	Liquid emissions from within the tree, leaking to the surface, often colonised by yeasts sometimes emerging under gaseous pressure.
fist – (see figure 48a)	A collection of knuckles.
former wood-pasture –	Wood-pasture that is no longer grazed.
forest –	A tract of land, usually owned by the Crown, and subject to special laws, mostly concerned with the preservation of game.
fused coppice –	A coppice stool with mature stems that have fused to give the appearance of a single stem.
girdling –	See ring-barking.
grazed high forest –	A form of wood-pasture where the trees are maidens; not coppice stools or pollards. The canopy is usually quite dense.
heartwood –	The dead, or predominantly dead, wood in the centre of tree species (eg oak) whose living sapwood has a determinate lifespan.
knuckle – (see figure 48a)	The top of the bole on a pollard. The point where branches have been repeatedly cut back which has become swollen.
hulk –	The remains of a large ancient tree, living or dead, with very little crown.
lapsed pollard –	A pollard that has not been cut for many years.
late-wood –	The wood produced by trees after the main flush of growth in the spring. The wood may have poorer water conduction ability than early-wood but is often more resistant to dysfunction and decay.
layering –	When aerial parts of a tree (or the whole tree) touch the ground and roots form new, but initially connected, plants. This can be natural or as a result of human influence.
leader –	The main (or topmost) shoot(s) of a tree.
lignin –	A component of wood cells that is cement-like and hard. The process of lignin deposition within the cellulose cell walls is called lignification.
lopping –	Cutting the lateral branches of a tree, but not the top.
maiden –	A tree that has not been modified by cutting. Unless it has been damaged by wind etc., it has its original natural crown.
mature –	A tree that has attained peak crown size and a shape that is different from the developing stage. The maximum point in the mature stage of a tree is also the pivotal point after which the ancient stage starts.
monolith –	A managed standing dead tree, usually with very few limbs.
mycorrhizal fungi –	Fungi forming an intimate and mutually beneficial association with the roots of trees.
natural pollard –	A tree that has been pollarded by 'natural' means, eg by wind or squirrel damage, sometimes also called a self-pollard. It will have originated from a maiden tree.

old growth woodland -	Woodland stands that have not been managed for over 200 years. Many of the trees have a large girth, and dead and dying trees are present.
old tree -	Any tree in the ancient phase.
open grown -	The form of trees grown in the absence of competition and shading which tends, in many species, to be squat and spreading.
over-mature -	A tree beyond full maturity. Usually used in commercial forestry to mean a tree past its commercial peak.
Park (or Historic Park) -	An area of enclosed land where domestic animals or, more usually, deer were, or are, kept among widely spaced trees. (Not a municipal park.)
parkland -	A form of wood-pasture where the trees are mostly open grown.
pasture woodland -	See wood-pasture.
pastured woodland -	A form of wood-pasture where the use by animals is periodic or seasonal, for example used as shelter in upland areas.
pathogen -	A micro-organism causing disease (adjective: pathogenic).
phloem -	The conductive tissue, composed of special cells, through which the products of photosynthesis are translocated throughout the tree.
phoenix regeneration -	A tree that has fallen or split apart that has successfully continued growing.
poll (v) and **polling -**	The formative process of removing the crown of a young maiden tree creating a pollard. Often now referred to as pollarding.
pollard (n) **-** (see figure 48a)	A tree cut once or repeatedly at a height above which grazing animals can reach the regenerating shoots. Usually cut on a semi-regular basis, with the whole or part of the crown removed.
pollard (v) **-**	The act of cutting an already created pollard. (Originally a noun derived from the transitive verb 'to poll', now used as a verb in its own right.)
pruning -	Cutting a tree with the intention of modifying its form or growth.
releasing (trees) -	Clearing competing younger trees from around a veteran.
relict pasture woodland -	Wood-pasture that is no longer grazed, also used for small remnants of a larger area of wood-pasture that are still grazed.
repollarding -	A confusing word, used in the past for both pollarding and restoration pollarding, best not used.
residual wood-pasture -	As former wood-pasture.
restoration pollarding -	The re-establishment of a cycle of pollarding on trees that have not been in a regular cycle for many years.
retrenchment -	A process whereby a tree with crown dieback forms a smaller, lower crown.
ring-barking -	The removal of a strip of inner and outer bark down to the cambium, all round the stem or trunk.
ripe wood -	Older wood in the centre of tree species (e.g. beech) where the sapwood gradually ages and is not converted to heartwood.
root spurs -	Equivalent to individual buttresses.
rot -	See decay.
sail area -	The amount of tree canopy that is exposed to the wind.
sap-flux -	See flux.
saprophyte -	As saprotrophic but pertaining to plants (adjective: saprophytic).
saprotrophe -	An organism obtaining nutrients from dead organic matter (adjective: saprotrophic).
sapwood -	The living xylem found in a woody plant. It either gradually loses viability over a number of years or is converted to a distinct heartwood which is largely dead.

saproxylic -	From the Greek Sapros (dead) and xylos (wood). Organisms that are dependent, during some part of their life cycle, upon wood or bark, usually dead or dying.
self-pollard -	See natural pollard.
singled coppice -	A coppice stool where all the limbs except one are removed.
shred -	A tree where the branches are periodically removed up the side of the trunk and then left to re-grow. A tuft is usually left at the top and occasionally an upper branch.
slime flux -	Flow of fluid out of the bark, partly due to microbial action.
snag -	In the USA a standing dead tree. Also used for a dead branch stub.
soft rot -	The wood decay that results from degradation of the cellulose in the cell walls, but without general erosion of the wall, by a fungus.
spalted -	A term used by wood workers to describe wood patterned by decay fungi.
springwood -	see earlywood.
stag headed -	The state of the crown of a tree when the dead branches protrude above the canopy. Often as a result of retrenchment and not a sign of ill health.
stool -	A tree that has been coppiced. Also used for that part of the tree which is retained after coppicing.
stored coppice -	A coppice stool retained beyond its normal coppice cycle.
stub -	A truncated limb on a tree, either cut or natural. Also used for a tree intermediate between a coppice stool and a pollard.
stow -	A pollard (local name).
sucker -	A shoot arising from a root.
summerwood -	see latewood.
timber -	Large tree trunks, suitable for sawing into planks.
topping -	cutting off most or all of the crown of a mature or semi-mature tree.
tree pool -	A pool of water on a tree, often where a branch or root joins the trunk. The bark may remain intact under the water or some decay may be taking place.
veteran tree -	A human valuation applied to trees in the ancient phase. The implication is that the tree has 'gone through the wars' and is a survivor.
water sprout/water shoot -	Epicormic growth.
white rot -	Decay where the lignin and cellulose are both broken down.
wildwood -	Woodland unaffected by neolithic or later civilisations.
withy -	Willow of 1-2 year old growth, used for baskets.
withy bed -	A group of willows used for withy.
wood -	Poles and branches of trees, smaller in diameter than timber.
wood mould -	The end result of the decay process, a rich, humus-like substance.
wood-pasture -	Land with trees that is grazed (often called pasture-woodland).
working tree -	A tree that is managed so that parts of it are used as a renewable resource for various purposes by man.
xylem -	Plant tissue that has the function of translocating water and mineral nutrients. In trees and shrubs the xylem is heavily lignified and has an additional function in providing structural support.

APPENDIX 1.

THE COMMON AND SCIENTIFIC NAMES OF TREES REFERRED TO IN THE TEXT

Alder		*Alnus glutinosa* (L.)
Apple		*Malus* Spp.
	Crab	*Malus sylvestris* (L.)
	Domestic	*Malus domestica* Borkh.
Ash		*Fraxinus excelsior* L.
Aspen		*Populus tremula* L.
Beech		*Fagus sylvatica* L.
Birch		*Betula* Spp.
	Downy	*Betula pubescens* Ehrh.
	Northern downy	*Betula pubescens* ssp. *tortuosa* (Ledeb.)
	Silver	*Betula pendula* Roth
Elm		*Ulmus* Spp.
	English	*Ulmus procera* Salisb.
	Wych	*Ulmus glabra* Hudson
Hawthorn		*Crataegus monogyna* Jacq.
Hazel		*Corylus avellana* L.
Holly		*Ilex aquifolium* L.
Hornbeam		*Carpinus betulus* L.
Horse chestnut		*Aesculus hippocastanum* L.
Lime		*Tilia* Spp.
	Small-leaved	*Tilia cordata* Miller
Maple		*Acer* Spp.
	Field	*Acer campestre* L.
	Norway	*Acer platanoides* L.
Oak		*Quercus* Spp.
	Pedunculate	*Quercus robur* L.
	Sessile	*Quercus petraea* (Mattuschka) Liebl.
Pear		*Pyrus* Spp.
Plane (London Plane)		*Platanus x hispanica* Miller ex Muenchh.
Poplar		*Populus* Spp.
	Black	*Populus nigra* ssp. *betulifolia* (Pursh)
Rowan		*Sorbus aucuparia* L.
Scots pine		*Pinus sylvestris* L.
Sweet chestnut		*Castanea sativa* Miller
Sycamore		*Acer pseudoplatanus* L.
Walnut		*Juglans regia* L.
Whitebeam		*Sorbus aria* (L.)
Wild Service		*Sorbus torminalis* (L.)
Willow		*Salix* Spp.
	Almond	*Salix triandra* L.
	Crack	*Salix fragilis* L.
	Osier	*Salix viminalis* L.
	Purple	*Salix purpurea* L.
	White	*Salix alba* L.
Yew		*Taxus baccata* L.

APPENDIX 2.

**LOWLAND WOOD-PASTURE AND PARKLAND:
A HABITAT ACTION PLAN**

1. Current status

1.1 Biological status

1.1.1 Lowland wood-pastures and parkland are the products of historic land management systems, and represent a vegetation structure rather than being a particular plant community. Typically this structure consists of large, open-grown or high forest trees (often pollards) at various densities, in a matrix of grazed grassland, heathland and/or woodland floras.

1.1.2 There are no reliable statistics on the extent of the overall resource, nor on historical and current rates of loss or degradation of this type of habitat. The figure of 10-20,000 ha "currently in a working condition" given in the 'habitat statement' of the UK Biodiversity Steering Group report is the current best estimate. This habitat is most common in southern Britain, but scattered examples occur throughout the country, for example Hamilton High Parks and Dalkeith Oakwood in Scotland. Outgrown wood-pasture and mature high forest remnants ('virgin forests') occur in northern and central Europe, but the number and continuity of ancient (veteran) trees with their associated distinctive saproxylic (wood-eating) fauna and epiphytic flora are more abundant in Britain than elsewhere. Parklands and wood-pasture may also be of interest for bats and birds and may preserve indigenous tree genotypes. These areas are outstanding at a European level.

1.1.3 These sites are frequently of national historic, cultural and landscape importance. Some, but not all, of the individual habitat components (lowland beech and yew woodland, lowland heathland, lowland dry acid grassland) are biodiversity action plan priority habitats in their own right. Requirements of these plans will need to be given due regard during implementation.

1.1.4 Included in this plan are:

i. Lowland wood-pastures and parklands derived from medieval forests and emparkments, wooded commons, parks and pastures with trees in them. Some have subsequently had a designed landscape superimposed in the 16th to 19th centuries. A range of native species usually predominates among the old trees but there may be non-native species that have been planted or regenerated naturally.

ii. Parklands with their origins in the 19th century or later where they contain much older trees derived from an earlier landscape.

iii. Under-managed and unmanaged wood-pastures with veteran trees, in a matrix of secondary woodland or scrub that has developed by regeneration and/or planting.

iv. Parkland or wood-pasture that has been converted to other land uses such as arable fields, forestry and amenity land, but where surviving veteran trees are of nature conservation interest. Some of the characteristic wood-pasture and parkland species may have survived this change in state.

1.1.5 Not included in this plan are:

i. Upland sheep-grazed closed-canopy oak woodland or Caledonian pine forest (see the respective plans for these habitats).

ii. Parklands with 19th century origins or later with none of the above characteristics.

1.1.6 In terms of the National Vegetation Classification (NVC) of plant communities lowland wood-pastures and parkland are most commonly associated with W10 *Quercus robur - Pteridium aquilinum - Rubus fruticosus* woodland, W14 *Fagus sylvatica - Rubus fruticosus* woodland, W15 *Fagus sylvatica - Deschampsia flexuosa* woodland and W16 *Quercus* spp. - *Betula* spp.- *Deschampsia flexuosa* woodland, although others may occur. In addition the more open wood-pastures and parkland may include various scrub, heathland, improved and unimproved grassland NVC communities.

1.2. Links with species action plans

1.2.1 Lowland wood-pasture and parkland is an important habitat for a number of priority species including violet click beetle *Limoniscus violaceu*s, the stag beetle *Lucanus cervus*, a bark beetle *Ernoporus tiliae*, a wood boring beetle *Gastrallus immarginatus*, orange-fruited elm lichen *Caloplaca luteoalba*, the lichens *Bacidia incompta*, *Enterographa sorediata* and *Schismatomma graphidioides*, the royal bolete fungi *Boletus regius*, oak polypore *Buglossoporus pulvinus* and the heart moth *Dicycla oo*. Their requirements should also be taken into account in the implementation of this plan. Other rare species include Moccas beetle *Hypebaeus flavipes*, and the lichen the New Forest parmelia *Parmelia minarium*.

2. Current factors affecting the habitat

2.1 Lack of younger generations of trees is producing a skewed age structure, leading to breaks in continuity of dead wood habitat and loss of specialised dependent species.

2.2 Neglect, and loss of expertise of traditional tree management techniques (eg pollarding) leading to trees collapsing or being felled for safety reasons.

2.3 Loss of veteran trees through disease (eg Dutch elm disease, oak dieback), physiological stress, such as drought and storm damage, and competition for resources with surrounding younger trees.

2.4 Removal of veteran trees and dead wood through perceptions of safety and tidiness where sites have high amenity use, forest hygiene, the supply of firewood or vandalism.

2.5 Damage to trees and roots from soil compaction and erosion caused by trampling by livestock and people and car parking.

2.6 Changes to ground-water levels leading to water stress and tree death, resulting from abstraction, drainage, neighbouring development, roads, prolonged drought and climate change.

2.7 Isolation and fragmentation of the remaining parklands and wood-pasture sites in the landscape. (Many of the species dependent on old trees are unable to move between these sites due to their poor powers of dispersal and the increasing distances they need to travel.)

2.8 Pasture loss through conversion to arable and other land-uses.

2.9 Pasture improvement through reseeding, deep ploughing, fertiliser and other chemical treatments, leading variously to tree root damage, loss of nectar-bearing plants, damage to the soil and epiphytes.

2.10 Inappropriate grazing levels: under-grazing leading to loss of habitat structure through bracken and scrub invasion; and over-grazing leading to bark browsing, soil compaction and loss of nectar plants.

VETERAN TREES
I N I T I A T I V E

2.11 Pollution derived either remotely from industry and traffic, or locally from agro-chemical application and nitrogen enrichment from pasture overstocking, causing damage to epiphyte communities and changes to soils.

3. Current action

3.1 Legal Status

3.1.1 For any woodland component of parkland and wood-pasture, national forestry policy includes a presumption against clearance of broad-leaved woodland for conversion to other land uses, and in particular seeks to maintain the special interest of ancient semi-natural woodland. Individual trees and groups may be afforded protection under the Town and Country Planning Act, 1990 and the Forestry Act, 1967. Felling licences from the Forestry Authority (FA) are normally required but veteran trees may be particularly at risk because fellings for safety reasons are exempt.

3.1.2 Statutory site protection plays an important part in the conservation of this habitat type. Designation as Sites of Special Scientific Interest (SSSI), or as Areas of Special Scientific Interest (ASSI) (Northern Ireland), of most larger areas of wood-pasture and parkland and most of the better-known sites of significance for invertebrates and lichens, ensures compulsory consultation with the statutory nature conservation agencies over management operations and development proposals. Designation under the EC Habitats Directive as Special Areas for Conservation will give additional protection to some parkland and wood-pasture sites. Some sites, including Moccas Park, Duncombe Park, Burnham Beeches, Leigh Woods, Hatfield Forest, parts of Bredon Hill, and Ashstead Common are also protected by National Nature Reserve (NNR) agreements.

3.1.3 Other sites receive some protection though initiatives such as the Inheritance Tax Exemption scheme or the declaration of National Trust and Corporation of London land properties as inalienable land. A few sites have specific legislation to protect them such as the Epping Forest Act of 1878.

3.1.4 The Moccas beetle *Hypebaeus flavipes*, violet click beetle *Limoniscus violaceus* and the orange-fruited elm lichen *Caloplaca luteoalba* and New Forest parmelia *Parmelia minarium* (all confined to parkland or wood-pasture) are fully protected under the 1981 Wildlife and Countryside Act, as are all species of bat and most tree-hole nesting birds. This Act also offers some protection to their "place of shelter".

3.1.5 There is recognition of the value of the habitat and individual old trees in various development plans, and landscape designations (eg by English Heritage, and CADW: Welsh Historic Monuments).

3.2 Management, research and guidance

3.2.1 There are a number of significant but currently uncoordinated inventories, datasets and registers of lowland wood-pasture and parkland. These include the Nature Conservancy Council's 1970s survey of parklands and wood-pastures of importance for the 'Mature Timber Habitat'; the Forestry Commission's National Inventory of Woodlands and Trees; The National Trust (NT) biological survey of NT-owned parkland and wood-pasture sites and English Nature's parkland inventory pilot study (1995) for Norfolk and Bedfordshire. English Heritage also has a register of parks and gardens, which is being upgraded between 1997 and 2000, and similar data for Wales is held by CADW: Welsh Historic Monuments. Scottish Natural Heritage maintains an inventory of Gardens and designed landscapes in Scotland. There is also an Inventory of Historic Parks and Gardens, based at University of York, which contains information on historically important sites and County Historic Gardens Trust data.

3.2.2 Surveys of saproxylic invertebrates and lichens have also been undertaken. These include the Countryside Council for Wales's strategic survey of Welsh parklands; K.N.A. Alexander's (National Trust) personal dataset on saproxylic beetle sites and the JNCC's Lower Plants and Invertebrate Site Registers. The British Lichen Society also maintains a database for parkland and wood-pasture.

3.2.3 Grant aid may be available for the management and restoration of parkland. The key sources of this aid include agri-environment schemes such as MAFF's Countryside Stewardship Scheme and the Countryside Council for Wales' Tir Cymen (which will be incorporated into an all-Wales Agri-environment scheme known as Tir Gofal in 1999) includes a scheme for Historic Landscapes and old orchards. Both of these schemes assist in the production of management plans, tree and grassland management and restoration of arable land to parkland. Other agri-environment schemes such as Environmentally Sensitive Areas (ESAs) and the Habitat Scheme (Wales) may subsidise the management or restoration of grassland and tree planting, and provide some protection for existing trees. The Forestry Authority's Woodland Grant Scheme is available for woodland with over 20% canopy cover.

3.2.4 The Veteran Trees Initiative, launched in 1996, aims to promote the value and importance of veteran trees and to conserve them wherever possible. This initiative is the result of a partnership between English Nature, English Heritage, the National Trust, Countryside Commission, Forest Authority, FRCA, Corporation of London and the Ancient Tree Forum. The initiative is developing a database for recording veteran trees, and provides advice on their management. It runs a national programme of demonstration and training days, and produces publications.

3.2.5 English Heritage's Conservation Area Partnerships, Scheduled Monuments and outstanding registered parklands initiative may also provide grant-aid and some Local Authority schemes, such as the Essex County Council's historic landscapes designation may also provide funding for management. The Countryside Council for Wales' "Orchards and Parklands Tree Scheme" grant aids management and restoration of parklands in Wales.

3.2.6 EC *Life* funding has also been awarded for management of the New Forest.

3.2.7 There is a wealth of information available from the Forestry Authority and other organisations and publications regarding all aspects of ancient woodland management. These include advice given locally through the statutory conservation agencies, the Farming and Wildlife Advisory Group, ADAS, the Countryside Advice and Information Service (Wales). The Forestry Commission's Arboricultural Advisory Service and English Heritage's Parks & Garden's Team of historians, landscape managers, ecologist and arboriculturalists can offer advice. The Ancient Tree Forum, an association of land managers, ecologists and arboriculturalists, provides advice, as do the voluntary and commercial sectors. The UK Forestry Standard and the Forestry Authority Guidelines for the management of semi-natural woodlands should be followed.

3.2.8 The British Lichen Society have produced a habitat management guide for lichens, including parklands and wood-pastures.

4. Action plan objectives and proposed targets

4.1 The objectives and targets cover habitat conservation, restoration and expansion. Key components include the need to secure favourable condition of key sites and, at appropriately targeted areas, to restore management or expand the habitat.

4.1.1 Protect and maintain the current extent (10-20,000 ha) and distribution of lowland wood-pasture and parkland in a favourable ecological condition.

4.1.2 Initiate in areas where examples of derelict wood-pasture and parkland occur a programme to restore 2,500 ha to favourable ecological condition by 2010.

4.1.3 By 2002 initiate the expansion of 500 ha of wood-pasture or parkland, in appropriate areas, to help reverse fragmentation and reduce the generation gap between veteran trees.

5. Proposed action with lead agencies

5.1 Policy and legislation

5.1.1 Implement the conclusions of the 1994 review of Tree Preservation Orders (TPO), including amendments to the Town and Country Planning Act 1990, to offer appropriate protection to veteran/dead trees. (Action: DETR)

5.1.2 Examine felling consent/licensing policy to consider whether additional protection for parkland, wood-pasture and individual veteran trees is needed. (Action: FA)

5.1.3 Examine whether improvements should be made in safety legislation, with respect to liability on owners in the event of injury or damage resulting from old trees, and its interpretation to reduce any unnecessary felling of trees on safety grounds. (Action: DETR, FA)

5.1.4 If Annex I of the EC Habitats Directive is revised ensure that it provides adequate coverage of UK parklands and wood-pasture habitats and species assemblages. (Action: DETR, JNCC)

5.1.5 When reviewing existing incentive schemes (eg Countryside Stewardship, Woodland Grant Scheme/ Woodland Improvement Grants, ESAs, Coed Cymru) attempt to ensure they enable and encourage the most appropriate management of parklands and wood-pasture, with their ancient trees. (Action: CCW, EN, FA, MAFF, SNH, SOAEFD, WOAD)

5.1.6 Promote modification of the Common Agricultural Policy to recognise and promote extensive pastoral systems, including wood-pasture. (Action: CCW, DETR, EN, MAFF, SNH, SOAEFD, WOAD).

5.1.7 Provide specific guidance about parklands, wood-pasture and individual veteran trees in Planning Policy Guidance notes (PPGs) by 2001. (Action: DETR, SNH, SOAEFD)

5.1.8 Review policy and practice regarding fencing of registered commons to allow reinstatement or control of grazing in wood-pasture commons, but without impediment to access by 2001. (Action: CC, DETR, FA, FE)

5.2 Site and safeguard and management

5.2.1 Ensure that SSSI coverage of important lowland wood-pasture and parkland sites is adequate through periodic review of the series. (Action: CCW, DETR, EN, SNH, SOAEFD, WO)

5.2.2 By 2004 designate those lowland wood-pasture sites approved by the EC as SACs under the Habitats Directive. (Action: CCW, DETR, EN, JNCC, SNH, SOAEFD, WO)

5.2.3 Encourage applications to buy and manage appropriate sites from potential funding sources. (Action: CC, CCW, EH, EN, SNH)

5.2.4 Encourage the development and implementation by 2004 of long-term integrated management plans for conservation and use of parklands and wood-pastures through agreements with site owners and in partnership with statutory wildlife, landscape and heritage agencies. (Action: CC, CCW, EN, FA, MAFF, SNH, SOAEFD, WOAD)

5.2.5 Promote re-establishment of grazing where appropriate in derelict wood-pasture and encourage the development of subsequent generations of veteran trees in all sites. (Action: CCW, EN, MAFF, SNH, SOAEFD, WOAD)

5.2.6 Promote the restoration of wood-pasture and parkland where old trees remain in former sites that are now arable fields or forestry plantations. (Action: CCW, FE, MAFF, WOAD)

5.2.7 By 2002 initiate programmes to expand parklands and wood-pasture sites in targeted areas. (Action: CC, CCW, EH, EN, FA, SNH)

5.2.8 Contribute to the implementation of relevant priority species action plans, through the integration of management requirements and advice, in conjunction with relevant steering groups. (Action: CCW, EN, MAFF, SNH, SOAEFD, WO)

5.2.9 Consider (re)establishment of key species dependent on veteran trees via translocation. (Action: CCW, EN, FA, FE, SNH)

5.3 Advisory

5.3.1 Develop a handbook(s) on best practice in management of parklands and wood-pasture in relation to wildlife, heritage and landscape conservation. (Action: CCW, DETR, EN, FA, SNH)

5.3.2 Develop clear guidance on safety-related risk assessment and reasonable practice, in conjunction with relevant landowners and management groups. (Action: DETR, FA).

5.3.3 Encourage training in best practice in park and wood-pasture management for site owners, site managers, land-agents, foresters, arboriculturalists and also for advisors and incentive scheme managers. (Action: CCW, EN, FA, MAFF, SNH)

5.4 International

5.4.1 Develop links with European organisations and programmes, such as the European Forestry Institute, the European Environment Agency and the European Centre for Nature Conservation to obtain estimates of the extent and distribution of comparable and related habitats, and exchange experience on research and management, by 2000. (Action: CCW, EN, FA, JNCC, SNH)

5.5 Monitoring and research

5.5.1 Produce a comprehensive list of all parkland and wood-pasture sites with pointers to other data sources and evaluations relating to both the natural and cultural heritage of each site, by 2002. Make this information available, through a data catalogue linked to the National Biodiversity Network. (Action: CC, CCW, EHS, EN, JNCC, SNH)

5.5.2 Develop and implement methods to assess the condition of wood-pastures and parkland by 2000 and encourage standardised recording and monitoring of tree population age structure,

survivorship and condition at key sites across the country in order to identify site specific and general trends. (Action: CCW, EHS, EN, FC, SNH)

5.5.3 Undertake a programme of targeted surveys of the biological interest of sites where lack of information is impeding their appropriate management, by 2005.

5.5.4 Ensure veteran tree recording is reflected in SSSI and Wildlife Site reporting and is input, as it becomes available, into local record centres as part of the National Biodiversity Network initiative. (Action: CCW, EN, FC, JNCC, SNH)

5.5.5 Develop and implement appropriate surveillance and monitoring programmes to assess progress towards action plan targets. (Action: CCW, EN, JNCC, SNH)

5.5.6 Encourage research into parkland and wood-pasture flora, including trees, and fauna in relation to tree and pasture management, including interactions and with invertebrates, fungi, soils, ground water levels and grazing animals and population dynamic studies. Ensure such research is coordinated with cultural heritage research. (Action: CCW, EH, EN, FC, SNH)

5.6 Communications and publicity

5.6.1 Increase awareness of the national and international importance and vulnerability of wood-pasture and parklands by promotional literature and events and encourage celebration of parkland and wood-pastures via the arts and media. (Action: CCW, EH, EN, SNH)

5.6.2 Increase awareness of the value in protecting veteran trees where these may be threatened by felling, for safety reasons, and promote alternative solutions such as pollarding or tree surgery. (Action: CCW, EHS, EN, FA, LA, SNH)

6. Costings

6.1 The successful implementation of the habitat action plans will have resource implications for both the private and public sectors. The data in the table below provide an estimate of the current expenditure on the habitat, primarily through agri-environment schemes and grant schemes, and the likely additional resource costs to the public and private sectors. These additional resource costs are based on the annual average over 5 and 10 years. The total expenditure for these periods of time is also given. Three-quarters of the additional resources are likely to fall to the public sector.

6.2 Current expenditure for the Woodland Grant Scheme has not been included as it was not possible to allocate expenditure to different woodland habitat types. It is estimated that 65 - 75% of the costs shown are additional to the current expenditure.

7. Key references

Department of the Environment, Transport and the Regions 1998. *Tree Preservation Orders Draft Regulations: a consultation paper.* DETR, London.

Forestry Authority 1994. Forestry Practice Guides: *The management of semi-natural woodlands.* Edinburgh: Forestry Authority.

Forestry Authority & Department of Agriculture for Northern Ireland 1998. *The UK Forestry Standard: the Government's approach to sustainable forestry.* Edinburgh Forestry Commission.

Harding, P.T., & Rose, F. 1986. *Pasture woodland in lowland England.* Huntingdon: Institute of Terrestrial Ecology.

Kirby K.J., Thomas, R. C., Key R.S., Mclean, I.F.G., & Hodgetts, N. 1995. Pasture woodland and its conservation in Britain. *Biological Journal of the Linnean Society,* **56** (Suppl.) 135-153.

Peterken, G.F. 1981. *Woodland conservation and management.* London: Chapman & Hall.

Ratcliffe, D.A. 1977. *A nature conservation review.* Cambridge: Cambridge University Press.

Rodwell, J.S. 1991. *British Plant Communities Volume 1: Woodlands and Scrub.* Cambridge University Press.

Costings for lowland wood-pasture and parkland

	Current expenditure	1st 5 yrs to 2003/2004	Next 10 yrs to 2013/2014
Current expenditure /£000/Yr	457.5		
Total average annual cost /£000/Yr		674.6	429.7
Total expenditure to 2004/£000		3373.0	
Total expenditure 2004 to 2014 /£000			4297.4

VETERAN TREES
I N I T I A T I V E

Appendix 3.

Historical information on pollarding

There are few written documents recording how and when trees were pollarded so the information is quite sparse. It has been supplemented to some degree by studies of tree rings. That which is available is summarised below:

1. Lop or fell wood in January. Leave a bough on the pollards and cut it away the following year. (T. Tusser 1573-1580).
2. Cut branches 1 - 2 ft (30 - 60 cm) from the body of the tree with a one handed axe, making sure that the bark is not damaged. Do not cut in sap time nor when the wind is in the north or east. (Fitzherbert 1523).
3. The commoners had rights to cut from the trees between All Saint's Day (1st November) and St. George's Day (23rd April). This was later changed to start on St. Martin's Day (11th November). Cuts were made with an axe. 'Commercial areas' assigned by the Lord of the Manor were cut between 1st February and 5th April. The trees were cut at intervals of 13-15 years. One area might have been cut on a more regular cycle of 10 years. Commoners cut below the previous point of cutting on hornbeam to maximise the amount of useable wood. Epping Forest, hornbeam and perhaps oak. (Dagley & Burman 1996).
4. An etching by Wenceslaus Hollar of Charles II shows an ?oak pollard recently cut with one large branch left as a sap riser (Falkus undated).
5. In the Lake District odd branches are removed from the pollards and fed to the sheep. Once be-barked the branches are used as fuel. (Quelch 1997).
6. Some limbs are left on the trees when they are cut in southern Europe. Some fodder trees were cut in the winter (e.g. Holly) others in the summer. (Green, 1996a).
7. Pollarding in Sweden was found to prevent the flowering of the trees because of the frequency of cutting. Shredding did not. (Andersen 1988).
8. Make the first cut when the tree is 25-35 years old. Then the intervals between cuts are 11-12 years for the first 4-5 cuts, getting gradually longer to 14 year intervals for cuts 6 and 7. Some branches were probably left on as sap lifters. Beech trees at Burnham Beeches. (Le Sueur 1931).
9. The usual period between cuts on pollards in Buckinghamshire was 7 years. (Le Sueur 1931).
10. A pollard oak in a park in Suffolk showed intervals between successive cuts to be 16, 14, 28, 24, 11 and 27 years (between 1602 and 1722). (Rackham 1988).
11. Cut at intervals of 12 years (though probably not on a regular cycle). Historical reference to Hatfield Forest (presumably the full range of species found in Hatfield Forest). (Rackham 1989).
12. An oak at Hatfield Forest was first cut at about 30 years old and then subsequently at intervals of 30, 12, 19, 17, 36 and 14 years. (Rackham 1989).
13. Another oak at Hatfield was first cut at 54 years of age and thereafter at intervals of 11-24 years. (Rackham 1989).
14. Cutting intervals were approximately 13 years in Epping Forest and 18-25 at Hainault Forest (presumably hornbeam and perhaps beech). (Rackham 1989). Oaks were cut at more or less the same rotation as coppice though there is no indication of a regular cycle and sometimes they were left much longer. (Rackham 1989).
15. Oaks in Kent were cut on a short rotation of 1-10 years to provide fodder and faggots for the salting industry. (D. Maylam pers. comm.)
16. By counting rings on pollard branches in Borrowdale it seems there was 30 years between each cutting on a tree but that selected branches were cut each time not all of them removed. (Mercer 1993).
17. In Kent, 'pegs' were always left when cutting pollards. (D. Maylam pers. comm.)

APPENDIX 4.

SPECIES SPECIFIC NOTES ON CUTTING VETERAN TREES AND YOUNG TREES IN ORDER TO CREATE POLLARDS FOR THE FUTURE.

These comments are based on personal experiences of many people, particularly members of the Ancient Tree Forum. Many relate to pollarding and it should be noted that they do not necessarily apply to all circumstances. **It is important also to read the notes in the main body of the text (especially chapter 4)** and to judge according to the situation. The response will also vary according to the local climate and conditions.

The species are listed in alphabetical order.

Alder
Veteran trees: Although alders are found along rivers and in damp areas throughout Britain, veterans and worked trees seem to be more localised. Alder wood is useful but the trees seem to have often been coppiced rather than pollarded. Since the leaves are unpalatable to stock the trees survived any grazing. Little work has been done on veteran alders but they are likely to respond in a similar way to willow, although perhaps a little less vigorous.
Young trees: Alder will respond well to cutting when young.

Ash
Veteran trees: Ash seems to be the ultimate unpredictable tree. Like oak it would be expected to respond well to cutting, but sometimes does not. Whilst there are a good number of veteran ash pollards in some areas of the countryside (eg East Anglia) there is not much of experience in recutting following a period of lapse. Removing all the branches from a pollard may delay regrowth until the end of the first growing season after cutting. Pollards with branches left on after cutting are more likely to be successful and cutting in a series of stages may be appropriate depending on the shape of the tree. Ash probably responds better in the north. Cutting veteran trees in the Lake District has been successful. In Sweden lapsed pollards are treated by removing the entire crown to 'shock' the tree to regrow and the results are good (Quelch pers. comm.).
Young trees: Young ash can be very late to break bud after the creation of new pollards and can take up to one year for this to happen. The sprouts tend to grow low on the trunk so cutting should be higher than the desired final height. Some losses are likely but may not be directly related to the diameter of the main stem. Jagged edged cuts are not necessary, simple ones and a complete removal of the crown is probably best. Cutting ash in late summer may be better than cutting in the winter. One method to try is to make an initial high cut and leave the tree for 4 to 5 years before cutting again, lower down, once sprouting has occurred (Figures 49 and 50).

Further reading: Mitchell (1989), White (1996) and Wisdom (1991).

Figure 49. *Cutting a maiden as tree in two stages.*

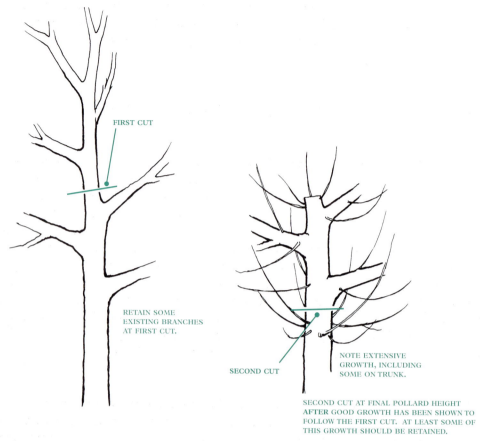

FIRST CUT

RETAIN SOME
EXISTING BRANCHES
AT FIRST CUT.

SECOND CUT

NOTE EXTENSIVE
GROWTH, INCLUDING
SOME ON TRUNK.

SECOND CUT AT FINAL POLLARD HEIGHT
AFTER GOOD GROWTH HAS BEEN SHOWN TO
FOLLOW THE FIRST CUT. AT LEAST SOME OF
THIS GROWTH SHOULD BE RETAINED.

Figure 50. *See colour plate page 93.*

Beech

Veteran trees: The chances of obtaining successful regrowth from an old beech pollards following complete removal of all branches are almost zero. Beech seems to be one of the most reluctant species to respond to cutting and, although it is reported to grow adventitiously from wound wood this does not seem to be the case with veteran trees in practice. The response of beech generally is to put on increased growth from the branches retained which may, over time, alter the shape of the tree quite considerably. While this growth does not result in 'new' branches as such, the life of the tree is extended and it seems likely that in most situations in the past this was how beech was cut. Rarely, epicormic growth results.

It has been recommended to leave a single leading branch on each tree (Mitchell 1989). On lapsed pollards (and old trees generally) this is almost certainly too little. Canopy reduction of 25 - 50% is probably more appropriate and more likely to succeed. Unfortunately, on many old beech trees the foliage is all very high up and it is not possible to retain the existing growth that is necessary for the survival of the tree when pollarding. In these instances there is very little which can be done. Unless action is required for safety reasons, these trees are probably best left unless growth can be stimulated lower on the branches first.

The responses of old trees to cutting seem to vary from site to site. This seems especially true of beech where similar types of cutting on trees in the same geographic area have produced very different responses. In the past it has been considered not worth trying to do anything with veteran beech. Recent work has shown that results can be good and in some situations it is worth trying. On some tree shapes it may be worth cutting in two or more stages over a period of about five years as described with oak; with other shapes this is unlikely to work. It is safest to assume that there will be no new growth on a branch at or below the point of cutting unless any is retained. Growth may be extensive from branches with retained foliage. (Figure 51).

Young trees: Even young beech must have more than one branch left on the stem. Growth occurs largely from the retained branches, it is far less likely from the main stem. Heavy losses occur unless branches are retained. Occasionally it is possible to remove all branches on some young trees without killing them but this is the exception rather than the rule.

Figure 51. *See colour plate page 94.*

Further reading: Dagley & Burman (1996), Read *et al.* (1991, 1996)

Birch

Though birch is not a tree found widely as an example of a pollard in Britain it was certainly frequently cut in the past in Scandinavia and northern Britain for fodder. For this reason the trees were probably mostly cut during summer months at intervals of five to seven years (Austad 1988). Young birch is likely to respond well to cutting and in general the whole crown can probably be removed though poor responses have been noted in maiden silver birch over 10 cm in diameter (N. Sanderson pers. comm.). Older trees have also been reported to die after removal of a high proportion of the crown in Sweden (Quelch pers. comm.) and veteran birch in the Lake District may show poor responses to surgery. Experiences with lopping large maiden birch trees in the south of England have also shown that they can be a little unpredictable, this may in part be related to their susceptibility to dry weather. The northern downy birch (*Betula pubescens* ssp. *tortuosa*) may be an easier species to work with than silver birch (N. Sanderson pers. comm., 1998a).

Black Poplar

Veteran trees: Black poplars are found close to water and in certain areas (eg the Vale of Aylesbury and East Anglia) are more widely distributed with some being boundary markers. Although poplars in general are not usually particularly long-lived trees, black poplars are perhaps the exception. Some have been pollarded in the past and this is likely to increase their chances of reaching veteran status. It is probably appropriate to treat them in a similar way to willows. It has been suggested that pollarding of black poplars is best done at the end of the growing season (mid to late summer) for the best chance of success although cutting in February is also likely to be successful. The responses following the cutting of lapsed pollards are variable. In some areas there has been only a 50% success rate so removal of the crown in stages may be more beneficial. Trees may also show good growth in the year following cutting and then die subsequently. Frequent pollarding of black poplar may not allow it to flower and seed.

Young trees: In general, species of poplar and are likely to be very responsive and easy to achieve good results with.

> ### *Black poplars in the Vale of Aylesbury*
>
> *A recent survey in the Vale of Aylesbury identified about 4,100 black poplars. Of the 3,660 found in the Buckinghamshire part of the Vale, 72% of the trees were distributed along streams and ditches and just 30 were females. The largest tree had a girth of 15.5 feet (472 cm) and the population as a whole was considered mature or over mature, with very low recruitment. Seventy three per cent of the trees were pollards and half of these were pollarded 'some time ago'. Aylesbury Countryside Management Project has repollarded over 50 trees in the last six years, with varying amounts of the crown removed. Virtually all survived and showed vigorous growth. The project is also taking cuttings so that plants of the same genetic stock can be perpetuated in the local area (ie on the same farms). A characteristic of the Aylesbury poplars is that they sucker well following cutting. The suckers can then help to form the next generation of trees too.*

Conifers - See Scots pine

Crab apple
Young trees: Crab apple generally seems to respond well to cutting.

Elm
Veteran trees: Due to Dutch elm disease it is unusual to find large trees in Great Britain now. Regularly pollarded elms existed in the past and could grow to large sizes (e.g. covering an area of 22 m x 22 m at Hailes Abbey, Gloucestershire, R. Finch pers. comm.). A few still remain in East Anglia where they were especially common in the past in village closes (small fields near to the settlement) (P. Harding pers. comm.). Wych elm has been shown to respond well to pollarding in Norway after a lapse of 40 or so years (Austad and Skogen 1990). It can be treated like lime but cutting was probably done mostly in the summer in the past as the leaves have a high nutritional value (Hauge 1988, Austad and Skogen 1990).

Young trees: Young elm responds well to cutting but losses are high due to Dutch elm disease.

Field Maple
Veteran trees: There are relatively few old field maple pollards. Those that have been cut in recent years seem to have grown well. One of 100 years of age, last cut 30 - 40 years ago grew well when stubs of 15 cm were left. A 350-year-old tree cut 20 years ago is growing well at Hatfield Forest (V. Forbes pers. comm.). Younger trees show an extremely good response so it is likely that older trees will grow well. Perhaps leave short stubs to be on the safe side.

Young trees: Field maple is likely to produce a very good response when cut, shoots may appear very low down on the trunk so it is necessary to cut higher. Bigger trees may show a poorer response in the first year than younger trees. Removing all the branches is probably best especially from burry trees.

Further reading: Wisdom 1991, Sisitka 1991a, 1991b.

Hawthorn
Veteran trees: Old trees seem to fall apart and layer themselves readily, e.g. at Hatfield (Sisitka 1991a, 1991b) but may not appear to be obvious pollards. When branches snap the regrowth is generally good. Cutting should probably be done as for field maple but may not be necessary if the trees behave as they do at Hatfield (ie falling apart but still growing). At other places (eg Croft) the trees grow differently and much more like typical pollards in shape. Due to their relatively light crowns there may be scope for cutting a few branches initially to check the response.

Young trees: Young hawthorn responds well to cuts. Winter cutting may be better than summer.

Hazel
Veteran trees: Veteran hazels are most likely to be coppice stools but in Scotland (eg Glen Finglas) single stemmed pollards are found. From experience with coppice, hazel will respond well to cutting as long as it has sufficient light.

Young trees: Young hazel responds well to cutting.

Holly
Veteran trees: Despite the nature of its leaves, holly has been cut in the past primarily as a fodder crop and seems to have been grown in several areas, eg south Pennines, the Marches counties and The New Forest for this reason. It was cut in the winter to provide extra feed for sheep or deer in times of severe weather. Old holly pollards still occur in Shropshire on the Stiperstones and the Hollies SSSI and a few trees have been cut in recent years. Some branches were left on these trees and those that were cut were not taken right back to the bolling, ie stubs were left. The work was done in the spring as holly is considered to be susceptible to frost.

Experience elsewhere on other species does not recommend spring cutting, though frost damage may be dependent on the situation. Despite its palatability, veteran holly appears to be able to withstand grazing pressure and in the New Forest the pollards seem to have been cut at a height lower than other species. It also seems very shade tolerant. Recent cutting of veteran trees in the New Forest, carried out in October and January with one or more branches left, are growing well. Cutting above the previous cut (i.e. in younger wood) may be more important than retaining a branch on veteran trees as older branches develop thicker bark (N. Sanderson pers. comm.). Some losses on maiden trees where no branches were retained suggest that it is prudent to retain branches.

Young trees: Young holly is generally very responsive to cutting (Figure 52). New holly pollards in the New Forest had a 88% success rate, but those failing as pollards mostly sprouted at the base (because they lack dormant buds?). Some losses do happen but holly usually occurs in such dense groups that a few deaths are not usually a problem. It may often be easier to cut a block of holly rather like a coppice plot, this ensures that there is enough light reaching the trees but may sometimes cause over exposure. In the New Forest blocks of 30 x 30 m were cut successfully. If only small numbers of the holly respond positively after cutting try leaving a branch or two on.

Figure 52. *See colour plate page 95.*

Further reading: Peterken *et al.* (1996), Radley (1961), Sanderson (1991), Spray (1981), Wall (1991).

Hornbeam

Veteran trees: Many hornbeam pollards in Eastern England have now been cut after years of neglect. In almost all instances the entire crown was removed. Results have generally been reasonable but the success rate is by no means 100% and at least at one site, trees flushed for several years and then declined.

Spring lopping is definitely deleterious for hornbeam. From the evidence of recent work on hornbeam it seems that lapsed pollards are likely to have a 70 - 90% success rate following complete removal of the crown. It is important to realise, however, that hornbeam does have a tendency to flush well initially and then die back. Trees in more open (less shaded) situations may be more susceptible to this. Success rates may be increased by leaving some existing growth on the trees, if they have any branches that are suitable, but this does not seem to have been proven. Leaving short stubs may be better than cutting right back to the bolling. It is recommended by some people to leave side branches and stubs of 2 - 3 m. Others suggest that leaving stubs on hornbeam is not necessary. In Epping Forest epicormic growth was produced from the base of the stubs or bolling and there was no difference between those trees with long stubs left and those with short ones. At Knebworth (Figure 53) no advantage was found in leaving long stubs.

Experiences include:
Hatfield Forest recorded a success rate of around 69% for trees cut in 1977, 1978 and 1979, during which the weather conditions were generally very dry (V. Forbes pers. comm.). Exposed trees may have dried out and then been attacked by the fungus *Bjerkandera adusta*. Had stubs been left the response might have been better.

In Hainault Forest and Epping Forest 80 - 91% success rates have been achieved. Trees at Hainault Forest also suffered from *Bjerkandera* and the vigour of trees seems to be not so good 5 years after cutting. At Epping Forest some standards were left to shade the pollards at least in some areas. At Gernon Bushes 95% success was obtained initially but some may not survive. Stubs of 60 cm left on veteran hornbeam in Kent resulted in good growth. (Hornbeam was reported to have originally been cut in Epping Forest using a long handled axe from the ground. It was also cut using a lopper whilst standing in the tree.)

At Knebworth 370 trees have been cut between 1991 and 1995 (a fifth each year) after a lapse in management of 55 years. Trees were cut in January and February and all branches were removed. A 93% survival rate was recorded with 60% of the trees showing strong regrowth. A greater chance of survival was recorded on trees that were 'intact', ie not hollow.

Young trees: Young hornbeam generally responds quite well and few/no losses seem to result when trees of up to 45 cm diameter are cut. It does seem able to survive in relatively poor light with all the branches removed (on trees up to 52 cm diameter) though leaving some short branches may be beneficial for bigger trees.

Figure 53. *See colour plate page 95.*

Further reading: Coombs (1991), Coop (1991), Dagley & Burman (1996), Rackham (1989), Sidwell (1996), Sisitka (1991a and 1991b), Warrington & Brookes (1998).

Horse Chestnut

Big horse chestnut trees are often a feature of designed landscapes. Their spectacular flowers in the spring led them to be widely planted near houses. Few pollards exist and they can become unstable, the branches falling out easily. However, in urban areas, trees cut regularly rarely break unless ingrown bark is present. The wood is brittle and difficult to work with but the trees generally respond quite well to pruning.

Lime

Veteran trees: Regeneration is likely to be very good; experiences include: a small-leaved lime 30 - 40 years since last cut which had every stem reduced to 0.3 m stubs, regeneration was prolific (Wisdom 1991). Street trees cut every other year with no stubs left responded well (Mayhew 1993). On street trees all the branches can be removed and there is no need to leave short stubs. However, for those of over about 40 years since the last cut it may be judicious to leave short stubs. Lime will take repeated pollarding on a short cycle, eg two to three years in a street situation. They may produce so much growth lower down with little at the top that they need trimming back on the trunk. Lime in Norway 40 years since last cut was left with 20 - 25 cm stubs, cut with chainsaws, and they regrew well (Austad and Skogen 1990).

Young trees: Lime is likely to be very responsive and easy to achieve good results with.

Oak

Veteran trees: As a general rule veteran oak trees respond to cutting more positively than beech but older trees do not respond nearly as well as young ones.

It is possible that regularly cut trees might have had all their branches removed though this is unlikely in certain parts of the country. When dealing with trees that have not been cut for many years it is best not to cut right back to the bolling. Although success can be obtained following total removal of the crown, the chance of regrowth is significantly higher when branches are retained and on most sites it is essential. It has proved beneficial in some cases to cut the tree in stages, first doing some initial crown reduction work and then coming back a few years later to reduce further. Oak regrowth is susceptible to mildew and this may be severe enough to affect the survival of the tree. Defoliating caterpillars may also have a detrimental effect.

Young trees: Creating young pollard oak trees seems to incur some losses (eg 17 out of 30 trees cut were lost at Hatfield, possibly due to a sudden exposure on release from the scrub or subsequent dry summers) not necessarily related to the diameter of the trunk. Cutting in two stages may help to reduce losses. Site variation can be quite considerable, eg Thorndon suffered heavy losses at one site when leaders were not left but few losses on another site when cutting was similar. Greater success rates are achieved in the wetter climates of the west. At Epping Forest and Hatch Park younger trees responded better and stubs of 30 cm also helped. When cutting older trees it may be better to cut in stages (see under ash below). Note that most work on oak has been on pedunculate oak. Few sessile trees have been cut.

Further reading: Dagley & Burman (1996), Sisitka (1991a, 1991b), Smith (1991).

Plane

As for lime (most being street trees). There may be some evidence that repeated cutting of London planes every 2-3 years may cause them to go into decline (eg after 55 years or so, J. White pers. comm.) if cut back to the same point since the tree might have 'run out' of dormant buds (Patch, 1991).

Scots pine (and other conifers)

As a general principle most conifers do not pollard (an exception being yew). It is extremely unusual to find old 'worked' conifers (however, Scots pine of over 5 m in girth and probably pollarded can be found at Glen Orchy (N. Sanderson pers. comm.)). Trees occur with forked stems (occasionally with more than two stems) due to damage early in life and exotic species in gardens are made into hedges but with old trees of this type there is little hope of expecting a positive response to any large-scale pruning. No regrowth can be expected to occur following such cutting though retained branches *may* grow towards the increased light levels if the tree is of an appropriate shape. Managing ancient Scots pine in the remnant Caledonian Forest is an issue not covered in this text.

Sorbus species (wild service, whitebeam, rowan)

Veteran trees: There is limited experience of working on veteran *Sorbus* trees and there are few pollards. Cutting may well elicit a good response.

Young trees: Cutting of young trees should produce a good response.

Sweet chestnut

Veteran trees: Big old sweet chestnuts seem to largely 'look after themselves'. Cutting young chestnut produces a very good response so cutting of older trees may not be problematic. On the other hand, chestnut behaves in many ways similar to oak and old oak trees can be rather unpredictable; thus it would be prudent to carry out some tentative work first to assess the response. It is probably appropriate not to cut the entire crown off and not to cut the branches flush.

Young trees: Sweet chestnut is likely to produce has a good response but cutting of the side branches may be needed to encourage growth towards the top of the tree.

Sycamore

Whilst there are examples of old sycamore pollards in Scotland these do not appear to have been cut again recently. It is likely that young trees will respond well to cutting.

Willow

Veteran trees: The oldest willows are likely to be pollards or their successors (ie bollings that have fallen apart or layered branches, see Figure 54). Large groups of trees are often a single clone of ancient origin.

The management of mature and veteran willows poses serious threats to the inexperienced or unsuspecting. The stresses and strains are not so obvious with willows as with other timber trees; the soft fibrous nature of the wood can often result in lengthy linear fractures when cutting is started. Tools used for cutting should always be in particularly good condition and a part sawn willow should never be left even momentarily. Because of the woolly nature of the saw-dust it is wise to have a coarse set on the saw. Veteran, neglected willows are liable to drop structurally unsound limbs readily and the bollings fall apart easily.

Crack willow, as its name suggests, is rather brittle and indeed can often be identified by the considerable accumulation of shed material around the base and lower branches caused by wind blow. Its growth form is also very poor, frequently developing low, heavy lateral limbs, often bowed and in any plane. On the upper surfaces organic material and moisture collect and epiphytic plants such as mosses and lichens grow. These all form slippery hazards for the tree surgeon. Other trees, shrubs and herbs are frequently found growing in the decomposing wood in the centre of pollards adding to the potential hazards.

There are 18 species of willow regarded as native to Britain. Most of these usually have multiple stems and are more like shrubs than trees. Both crack willow and white willow have more tendency to grow with a single trunk and have been regularly pollarded in the past, often for fencing. Almond willow, purple willow and the common osier were the species most frequently coppiced in withy beds or willow holts. The growth is rapid, over 3 m per year, and the pliable rods were used for basketry, cart bottoms, hurdles and other rural crafts. There are also examples of common osiers in Suffolk cropped as shreds from saplings or singled, grown out coppice (P. Read pers. comm.).

Willow is one type of tree where pollarding has more or less continued from the 13th century to the present day. They are not especially long-lived trees and many pollards with ancient characteristics found along riversides may well have been created in relatively recent times. Today well managed pollards help stabilise the ground alongside rivers but trees cut on the river-side only, as frequently happens, produces lop sided trees and can cause problems.

It is reputed that willow pollards are best cut in February but most species will respond well when cut at any time of the year. Late summer cuts are less advisable as the young growth may suffer from winter weather conditions. Early spring growth was a convenient form of forage after winter food shortages.

Old willow pollards are usually able to respond well to complete removal of the canopy with minimal stubs left, especially if it has not been many years since the last cut. However, if there is a small amount of leafy material very close to the bolling on short stems it is worth leaving this. It has been suggested that leaving a single larger branch on willow pollards may be detrimental because the tree does not produce a proper flush of new stems from the bolling and the branch often breaks out if one is left isolated (J. White pers. comm.). Also, in crack willows, the brittle nature of the wood often results in at best the branch breaking in the wind and at worst some damage to the trunk of the tree. At the Nene Park Trust near Peterborough it was found that removing the whole crown of old lapsed white willow pollards did not result in such good growth as was expected. Some trees died and in others some partial death of the bole occurred, though this was often obscured by the good growth from the living parts of the tree. To overcome this, the trees are now cut in a series of stages over 3 or 4 years. At the first cut plenty of wood is retained on each major limb and this results in good growth. In the following years the tree is cut back further until just 15 - 30 cm of stem is left on the bole. This method is proving very successful and younger willows are also now cut in two stages.

Old willows that are not pollards may not respond so well to removing the entire crown and careful reduction work may be more appropriate. However, white willows of a variety of ages in Suffolk have grown well following complete decapitation (P. Read pers. comm.).

One important point when cutting willows (as with all tree species) is to try to avoid cutting all the trees in a group at the same time. Because they tend to occur in small clumps they are often all pollarded together for financial and practical reasons. While this is not a problem for the trees it can be for the populations of invertebrates requiring a particular stage of growth for their livelihood. Willow branches that are horizontal or have fallen are especially valuable for bryophytes and should be retained if possible. It is increasingly being realised that willows make an outstanding contribution to biodiversity in the north of the country and may exceed that of oak (N. Lewis pers. comm.) so these aspects are important to consider.

Young trees: Young willow should respond well. New trees were often started just by placing a newly cut pole in the ground to root and then the new branches were cut back to the height of the pole each time.

Further reading: Edlin 1956, Rackham 1986, 1990, Braun & Konold 1998.

Figure 54. *See colour plate page 96.*

Yew

Veteran trees: Yews are probably the longest lived trees found naturally in Britain. The trees are usually able to survive quite happily with minimum intervention but pruning of old trees is occasionally necessary. Yew does not usually occur as pollards but pruning of some or all the branches usually produces a good response. Over 50 trees at Westonbirt have been cut to a bare stump over the past 20 years with no losses (J. White pers. comm.). However, pruning of branch tips (eg along roadsides) in Herefordshire has resulted in die back along some branches treated (H. Stace pers. comm.). Yew is extremely shade tolerant and some individuals have abundant epicormic growth.

Young trees: Yew generally responds very well to cutting at any age.

VETERAN TREES
I N I T I A T I V E

APPENDIX 5.

CALCULATING THE AMOUNT OF DEAD WOOD WITHIN A WOODLAND

The amount of dead wood is recorded by line-intercept sampling. It is recommended that five to ten transects are undertaken in each block of woodland to be sampled. The transects can be 25 m or 50 m, depending on the amount of dead wood available.

For each transect:
1. Arrange a starting point and direction for each transect beforehand or establish them randomly within the plot.
2. Lay down a tape (or rope) along the line of the transect.
3. Record each log or piece of wood (more than 5 cm diameter at the point of intersection) that the tape crosses. For each log note its diameter at intersection.
4. Record any standing dead trees with their centre within 2 m either side of the transect line. Note the diameter at breast height (1.3 m) for any more than 5 cm diameter.
5. The length of fallen logs in the stand can be estimated using the following equation:

$$L = \pi \, 10^4 \, N \, (2t)^{-1}$$

Where N is the number of intersections, t is the transect length (in metres) and L is the total length of fallen wood per hectare (in metres). The conversion factor of 10^4 is used to convert the results into metres per hectare.

6. The next step is to estimate the volume of fallen logs. To do this, use the diameter of the logs at the point of intersection and assign them to diameter classes, eg 5 - 10 cm, 11 - 20 cm, 21 -30 cm, 31 - 40 cm and >40 cm.
7. Calculate the mean cross sectional area for each size class and the length of fallen logs in that size class (using the above equation).
8. The total volume for each size class is estimated by:

$$V = nd^2 \, \pi^2 \, 10^4 \, (8t)^{-1}$$

Where V is the total volume of fallen logs of diameter class d, n is the number of intersections for logs of diameter d, and t is the total length of transect as used before.
9. The total volume for the stand is the sum of the volumes for each diameter class.

The following benchmarks can be used to give a comparison of amounts of fallen dead wood (according to Kirby et al. (1998)):

Level of dead wood	Volume of fallen wood (m³ha⁻¹)	No. of standing dead trees (ha⁻¹)	Size distribution of standing dead trees
Low	<20	0 - 10	All <10cm dbh
Medium	20 - 40	11 - 50	Some >10cm dbh
High	>40	>50	Some >40cm dbh

VETERAN TREES
INITIATIVE

APPENDIX 6.

ORGANISATIONS GIVING ADVICE AND/OR GRANT AID★ ON VETERAN TREES OR ASPECTS OF VETERAN TREES

MOST OF THE FOLLOWING ORGANISATIONS ARE NOT ABLE TO GIVE COMPREHENSIVE ADVICE ON VETERAN TREES

Ancient Tree Forum
P.O. Box 49
Ashtead
Surrey
KT21 1YG

Arboricultural Advisory and Information Service (Tree Advice Trust)
Alice Holt Lodge
Wrecclesham
Farnham
Surrey
GU10 4LH
Tel: 0897 161 147
(Premium rate line for technical advice)
01420 22022 (For publications, subscriptions etc.)
Fax: 01420 220 000

Arboricultural Association
Ampfield House
Ampfield
Romsey
Hampshire
SO51 9PA
Tel: 01794 368 717
Fax: 01794 368 978

Bat Conservation Trust
15 Cloisters House
8 Battersea Park Road
London
SW8 4BG
Tel: 0207 627 2629
Bat Helpline: 020 7627 8822

CADW: Welsh Historic Monuments
Crown Buildings
Cathays Park
Cardiff
CF1 3NQ
Tel: 02920 500 200

Countryside Agency
John Dower House
Crescent Place
Cheltenham
Gloucestershire
GL50 3RA
Tel: 01242 521 381

Countryside Council for Wales
(SNCO for Wales)
Plas Penrhos
Fford Penrhos
Bangor Gwynedd
LL57 2LQ
Tel: 01248 385500

For **Tir Gofal** contact
Countryside Council for Wales
1st floor
Ladywell House
Park Street
Newtown Powys
SY15 1RD
Tel: 01686 613400

DETR
(for Biodiversity Action Plans etc.)
Room 902
Tollgate House
Houlton Street
Bristol
BS2 9DJ
Tel: 01179 876154

DETR
(for Tree Preservation Orders)
Rural Development Division 3/B5
Eland House
Bressenden Place
London
SW1E 5DU
Tel: 020 7890 5623

English Heritage
23 Savile Row
London
W1X 1AB
Tel: 020 7973 3000

English Nature
(SNCO for England)
Northminster House
Peterborough
Cambridgeshire
PE1 1UA
Tel: 01733 455 000
Enquiry Service (for veteran tree recording form and publications):
01733 455101

Environment and Heritage Service
(SNCO for Northern Ireland)
Commonwealth House
35 Castle Street
Belfast
BT1 1GU
Tel: 02890 251 477

★Forestry Commission
(Woodland Grant Scheme)
National Office for England
Great Eastern House
Tenison Road
Cambridge
CB1 2DU
Tel: 01223 314 546

★FRCA (Countryside Stewardship and Environmentally Sensitive Area payments).
Contact your nearest regional office, or phone: 0645 335 577.

Historic Scotland
Longmore House
Salisbury Place
Edinburgh
EH9 1SH
Tel: 0131 668 8600
Fax: 0131 668 8789

★Heritage Lottery Fund
7 Holbein Place
London
SW1W 8NR
Tel: 020 7591 6041 (information and publications)
Tel: 020 7591 6042/3/4/5

Scottish Natural Heritage
(SNCO for Scotland)
2 Anderson Place
Edinburgh
EH6 5NP
Tel: 0131 554 9797

Trees of Time and Place
Trees of Time and Place
Co-ordinator
c/o ESSO UK PLC
Mailpoint 08
ESSO House
Ermyn Way
Leatherhead
Surrey
KT22 8UX
Tel: 01372 222528
Fax: 01372 223222

Tree Register of the British Isles (TROBI)
Secretary
77a Hall End
Wootton
Bedfordshire
MK43 9HP
Tel: 01234 768 884

Woodland Trust
Autumn Park
Dysart Road
Grantham
NG31 6LL
Tel: 01476 581111
Fax: 01476 590808

APPENDIX 7.

ABBREVIATIONS USED IN THE TEXT

ASSI — Areas of Scientific Interest

ATF — Ancient Tree Forum

CA — Countryside Agency

CADW — Welsh Historic Monuments

CCW — Countryside Council for Wales

DETR — Department of Environment, Transport and the Regions

EH — English Heritage

EN — English Nature

FE — Forest Enterprise

FA — Forestry Authority

FC — Forestry Commission

HMSO — Her Majesty's Stationery Office

JNCC — Joint Nature Conservation Committee

FWAG — Farming and Wildlife Advisory Group

MAFF — Ministry of Agriculture, Fisheries and Food

NNR — National Nature Reserve

RSPB — The Royal Society for the Protection of Birds

SAC — Special Area of Conservation

SNCO — Statutory Nature Conservation Organisation

SNH — Scottish Natural Heritage

SOAFED — Scottish Office Agriculture, Environment and Fisheries Department

SSSI — Site of Special Scientific Interest

TPO — Tree Preservation Order

VTI — Veteran Trees Initiative

WATCH — Junior branch of The Wildlife Trust

WOAD — Welsh Office Agriculture Department

References and Further Reading

ADAMS, K.J. 1996. The bryophyte flora of pollards and pollarded woodland with particular reference to eastern England. *In: Pollard and veteran tree management II;* ed. by H.J. READ, 12-16, Corporation of London.

ALEXANDER, K., & GREEN, T. 1993. Dead wood - eyesore or ecosystem. *Enact* **1**(1): 11-14.

ALEXANDER, K., GREEN, T., & KEY, R. 1996. The management of overmature tree populations for nature conservation - the basic guidelines. *In: Pollard and veteran tree management II;* ed. by H.J. READ, 122-135, Corporation of London.

ALEXANDER, K., GREEN, T., & KEY, R.1998. Managing our ancient trees. *Tree News,* Spring 1998: 10-13.

ANDERSEN, S.T. 1988. Changes in agricultural practices in the Holocene indicated in pollen diagram from a small hollow in Denmark. *In:* H.H. BIRKS, H.J. BIRKS, P.E. KALAND and D. MOE, eds. *The cultural landscape - Past, present and future;* 395-407 Cambridge University Press.

ANON 1996. Exploding beeches. *Tree News,* Spring 1996: 5.

ATKINSON, M. 1996. Creating new pollards at Hatfield Forest, Essex. *In: Pollard and veteran tree management II;* ed. by H.J. READ, 86-88, Corporation of London.

AUSTAD, I. 1988. Tree pollarding in Western Norway. *In:* H.H. BIRKS, H.J. BIRKS, P.E. KALAND and D. MOE eds. *The cultural landscape - Past, present and future;* 11-29, Cambridge University Press.

AUSTAD, I., & Skogen, A. 1990. Restoration of a deciduous woodland in Western Norway formerly used for fodder production: Effects on tree canopy and field layer. *Vegetatio,* **88**: 1-20.

BACON, J. 1994. A prickly problem. *Enact,* **2**(1): 12-15.

BACON, J. 1995. Removing the prickles. *Enact,* **3**(2): 10-11.

BACON, J., & Overbury, T. 1998. Pulling tall weeds. *Enact,* **6**(2): 7-9.

BARWICK, P. (1996). The Birklands oak project. *In: Pollard and veteran tree management II;* ed. by H.J. READ, 69-70, Corporation of London.

BAT CONSERVATION TRUST 1997. *Bats and trees.* Bat Conservation Trust.

BATTELL, G. 1996. Our Ancient Trees - The Way Ahead. *In: Pollard and veteran tree management II;* ed. by H.J. READ, 140, Corporation of London.

BAXTER, T. 1992. *The eternal yew.* The Self Publishing Association Ltd.

BECKETT, K.A. 1975. *The love of trees.* London: Octopus Books.

BERGENDORFF, C. & EMANUELSSON, U. 1996. History and traces of coppicing and pollarding in Scania, South Sweden. *In:* H. SLOTTE, and H. Göransson, eds. *Lövtäkt ochstubbskottsbruk II.* 235-304, Kungl. Skogs-och lantbruksakademien. Stockholm.

BODDY, L., & RAYNER, A.D.M. 1983. Origins of decay in living deciduous trees: The role of moisture content and a re-appraisal of the expanded concept of tree decay. *New Phytologist,* **94**: 623-641.

BOWES, B.G. 1996. Variations in the form of beech trees in Scotland. *In: Pollard and veteran tree management II*; ed. by H.J. READ, 77-81, Corporation of London.

BRAUN, B., & KONOLD, W. 1998. *Kopfweiden.* Beih. Veröff. Naturschutz Landschaftspflege Bad.-Württ **89**: 1-240. Karlsruhe.

BRITISH TRUST FOR CONSERVATION VOLUNTEERS 1980. *Woodlands - A practical Handbook.* BTCV.

BUCKLEY, G.P. 1992. *Ecology and management of coppice woodland.* London. Chapman & Hall.

BULLOCK, D.J., & Alexander, K. 1998. *Parklands - the way forward.* English Nature Research Report No. 295. Peterborough.

BURGESS, N.D., & Evans, C. 1989. *Management techniques for the control of Bracken.* RSPB.

BURMAN, P. 1991. Pollarding at Epping Forest. *In: Pollard and veteran tree management*; ed. by H.J. READ, 42-43. Corporation of London.

BUTTERFLY CONSERVATION 1998. *Bracken for butterflies.* Butterfly Conservation.

CIMON, N. 1983. *A simple model to predict snag levels in managed forests.* Snag habitat symposium 7-9 June 1983, Flagstaff, Arizona.

CLARKE, A. 1992. The effect of sunlight on the renovation of ancient neglected pollards. Unpublished project submitted towards HND in Rural Resource Management. Seale Hayne.

CLAYDEN, D. 1996. Data collection and analysis of veteran tree populations: a plea for co-ordination - with an example from Duncombe Park NNR/SSSI, North Yorkshire. *In: Pollard and veteran tree management II*; ed. by H.J. READ, 55-60, Corporation of London.

CODER, K. 1996. What was old is new again. *Arborist News,* August 1996: 53-59.

COLE, W. 1894. The management of Epping Forest: Memorials to the Committee and second official report of the experts. *Essex Naturalist,* **9**: 74-80.

COLEMAN, N. 1996. Maiden pollarding at Thorndon Country Park. *In: Pollard and veteran tree management II*; ed. by H.J. READ, 89-90, Corporation of London.

CONSERVATION MANAGEMENT SYSTEM 1996. *Conservation management system.* CMS partnership.

COOKE, A.S. 1997. *Avermectin use in livestock.* FWAG information leaflet.

COOMBES, N. 1991. Notes on re-pollarding hornbeam at Gernon Bushes Nature Reserve. *In: Pollard and veteran tree management*; ed. by H.J. READ, 49. Corporation of London.

COOP, G. 1991. Hainault Forest Country Park. *In: Pollard and veteran tree management*; ed. by H.J. READ, 44-45. Corporation of London.

COUNTRYSIDE COMMISSION 1998. *Site Management planning.* CCP 527.

VETERAN TREES
I N I T I A T I V E

CROFTS, A., & JEFFERSON, R.G. 1999. *The lowland grassland management handbook* 2nd edition. Peterborough: English Nature/The Wildlife Trusts..

DAGLEY, J., & BURMAN, P. 1996. The management of the pollards of Epping Forest: Its history and revival. *In: Pollard and veteran tree management II;* ed. by H.J. READ, 29-41, Corporation of London.

DAMANT, C. 1996. Possible bundle planting in Buckinghamshire's Chilterns. *In: Pollard and veteran tree management II;* ed. by H.J. READ, 93-97, Corporation of London.

DEBOIS LANDSCAPE SURVEY GROUP 1997. Designed landscapes in Scotland: Notes on their planting and management. Report for Scottish Natural Heritage. Contract SNH/061/94/LRB

DOLWIN, J.A., LONSDALE, D., & BARNETT, J. (1998). Detection of decay in trees. Arboricultural Research and Information Note 144-98-EXT.

DOUGALL, M., & DICKSON, J. 1997. Old managed oaks in the Glasgow area. *In: Scottish woodland history,* ed. by T.C. SMOUT, Edinburgh: Scottish Cultural Press.

EDLIN, H.L. 1956. *Trees, woods and man.* London: Collins, New Naturalist.

EDLIN, H.L. 1971. Woodland notebook: Good bye to the pollards. *Quarterly Journal of Forestry,* LXV: 157-165.

EMANUELSSON, U. 1988. A model for describing the development of the cultural landscape. *In: The cultural Landscape - Past, present and future,* eds. by H.H. BIRKS, H.J. BIRKS, P.E. KALAND and D. MOE, 111-121, Cambridge University Press.

ENGLISH HERITAGE 1998. *The register of parks and gardens of special historic interest.* English Heritage.

ENGLISH NATURE 1994. *Species conservation handbook.* Peterborough: English Nature.

ENGLISH NATURE 1996. *Guide to the care of ancient trees.* Peterborough: English Nature.

ENGLISH NATURE 1998. *The conservation of wild mushrooms.* Peterborough: English Nature.

FALKUS, C. undated. *The life and times of Charles II.* London: Weidenfeld and Nicholson.

FAY, N. 1996. Recording veteran trees. *In: Pollard and veteran tree management II;* ed. by H.J. READ, 136-137, Corporation of London.

FERRIS-KAAN, R., LONSDALE, D., & WINTER, T. 1993. *The conservation management of deadwood in trees.* Forestry Authority, Research Information Note: 241.

FINCH, R. 1996. An alternative method of crown reduction for ancient pollards and dead trees. *In: Pollard and veteran tree management II;* ed. by H.J. READ, 98-99, Corporation of London.

FITZHERBERT 1523. *Art of husbandrye.* London.

FORBES, V., & WARNOCK, B. 1996. Ashtead Common: A case study in conserving a forest of veteran trees. *In: Pollard and veteran tree management II;* ed. by H.J. READ, 61-64, Corporation of London.

FORESTRY AUTHORITY 1998. *The U.K. forestry standard.* Edinburgh: Forestry Authority.

FORESTRY COMMISSION 1990. *Forest nature conservation guidelines.* London: HMSO.

FORESTRY COMMISSION 1997. *Tree felling - getting permission.* Forestry Commission.

FOWLES, A.P. 1997. The saproxylic quality index: An evaluation of dead wood habitats based on rarity scores with examples from Wales. *Coleopterist,* **6**: 61-66.

FOWLES, A.P., ALEXANDER, K.N.A., & KEY, R.S. 1999. The saproxylic quality index: Evaluating wooded habitats for the conservation of dead wood coleoptera. *Coleopterist* (in press).

FRETWELL, K. 1996. Bundle debate. Letter. *Tree News,* Spring 1996:

FRETWELL, K., & GREEN, E.E. 1996. Methods for tagging trees. *In: Pollard and veteran tree management II;* ed. by H.J. READ, 138-139, Corporation of London.

FRY, R., & LONSDALE, D. 1991. *Habitat conservation for insects - a neglected green issue.* Middx: The Amateur Entomologists' Society.

FULLER, R.J. 1995. *Birdlife of woodland and forest.* Cambridge: Cambridge University Press.

FULLER, R.J., & WARREN, M.S. 1993. *Coppiced woodlands: their management for wildlife.* Peterborough: JNCC.

GARDEN HISTORY SOCIETY 1992. Nature conservation and historic parks. Unpublished manuscript.

GIBSON, C. 1997. Reintroducing stock grazing to Savernake Forest: A feasibility study. English Nature Research Reports. No. 224. Peterborough: English Nature.

GILBERT, O.L. 1984. Some effects of disturbance on the lichen flora of oceanic hazel woodland. Lichenologist 16: 21-30.

GILBERT, O.L. 1991. A successful transplant operation involving *Lobaria amplissima. Lichenologist,* **23**: 73-76.

GRAHAM, M. undated. Trees for life - A guide to ancient trees. (leaflet) Leicester.

GREEN, T. 1991. Simply fungi. *In: Pollard and veteran tree management;* ed. by H.J. READ, 26-27. Corporation of London.

GREEN, T. 1993. Introducing new theories on wood rotting fungi. *Mycologist,* **11**(2): 80-84.

GREEN, T. 1994. Woodman and the working tree. *Arboricultural Journal,* **18**: 205-207.

GREEN, T. 1995a. Creating decaying trees. *British Wildlife,* **6**(5): 310-311.

GREEN, T. 1995b. Nest boxes. *British Wildlife,* **6**(6): 378.

GREEN, T. 1995c. Advantages of bundle planting. *Tree News,* Spring/Summer 1995: 19-20.

GREEN, E.E. 1996a. Thoughts on pollarding. *In: Pollard and veteran tree management II;* ed. by H.J. READ, 1-5, Corporation of London.

GREEN, E.E. 1996b. Bundle planting. *In: Pollard and veteran tree management II;* ed. by H.J. READ, 91-92, Corporation of London.

GREEN, T. 1996c. Deadwood for wildlife. *Enact,* **4**(1): 10-11.

GREEN, T. 1996d. Woodpecker and bat trees. *British Wildlife,* **8**(1): 42-43.

GREEN, T. 1996e. Pollarding - origins and some practical advice. *British Wildlife,* **8**(2): 100-105.

GREEN, T. 1997. Maintaining dead wood. *British Wildlife,* **9**(2): 111.

HÆGGSTRÖM, C-A. 1992. Wooded meadows and the use of deciduous trees for fodder, fuel, carpentry and building purposes. *Protoindustries et histoire des forêts.* 1992 (3): 151-162.

HÆGGSTRÖM, C-A. 1994. Pollards in Art. *Botanical Journal of Scotland,* **46**: 682-687.

HÆGGSTRÖM, C-A. 1995. Lövängar I Norden och Balticum. *Nordenskiöld-Samfundets tidskrift,* **54**: 21-58.

HÆGGSTRÖM, C-A. 1998. Pollard meadows: multiple use of human-made nature. In: ed. The ecological history of European forests, eds. by K.J. KIRBY and C. WATKINS, CAB international: 33-41.

HAMMOND, P.M., & HARDING, P.T. 1991. Saproxylic invertebrate assemblages in British woodlands: Their conservation significance and its evaluation. *In: Pollard and veteran tree management;* ed. by H.J. READ, 30-37. Corporation of London.

HAMPSHIRE COUNTY COUNCIL 1991. *Hazel coppice.* Hampshire County Council.

HARDING, P.T., & ALEXANDER, K.N.A. 1993. The saproxylic invertebrates of historic parklands: Progress & problems. *In: Dead wood matters: the ecology and conservation of saproxylic invertebrates in Britain,* eds. by K.J. KIRBY and C.M. DRAKE, English Nature Science No. 7.

HARDING, P.T., ALEXANDER, K.N.A., ANDERSON, M.A., & LONSDALE, D. 1988. *Conserving insect habitats provided in dead broadleaved wood by the wind damage of 16th October 1987.* Research Information Note 136. Forestry Commission Research Division.

HARDING, P.T., & ROSE, F. 1986. *Pasture woodland in lowland Britain.* Huntingdon: Institute of Terrestrial Ecology.

HARDING, P.T., & WALL, T. in press. *Moccas: An English deer park.* Peterborough: English Nature.

HAUGE, L. 1988. Galdane, Lærdal, Western Norway - Management and restoration of the cultural landscape. *In: The cultural landscape - past, present and future;* eds. by H.H. BIRKS, H.J. BIRKS, P.E. KALAND and D. MOE, 31-45, Cambridge University Press.

HAYWARD, N. 1996. Conservation and safety: the beginnings of a veteran tree management strategy for the New Forest. *In: Pollard and veteran tree management II;* ed. by H.J. READ, 71-74, Corporation of London.

HODGE, S.J., & PETERKEN, G.F. 1998. Deadwood in British forests: priorities and a strategy. *Forestry*, **71**(2): 99-112.

HODGETTS, N. 1989. *Parkland management for lichens.* CSD Notes No. 48. Peterborough: Nature Conservancy Council.

HOLMES, M. 1996. Ancient trees - their importance to bats. *In: Pollard and veteran tree management II;* ed. by H.J. READ, 19-20, Corporation of London.

HOLMES, M. 1997. Bats and trees. *Tree News*, Autumn 1997: 16-17.

HOLMES, M. 1998. Managing woods for bats. *Enact*, **6**(4): 8-10.

HOPKINS, E. 1998. *Trees and bats.* Arboricultural Association Guidance Note 1. 36pp.

ING, B. 1996. The importance of ancient woodlands in the conservation of larger fungi. *In: Pollard and veteran tree management II;* ed. by H.J. READ, 10-11, Corporation of London.

JNCC. undated. *Chemical alternatives to treatment of cattle with ivermectin.* Peterborough: JNCC.

KERR, G. 1992. Tree shelters: Uses and abuses. *Tree News*, February 1992: 16-17.

KEY, R. 1991. Guide for the investigation of dead wood. In: A Coleopterists handbook, ed. by J. COOTER *et al.* 3rd edition. Feltham: Amateur Entomologists' Society.

KEY, R. 1993. What are saproxylic invertebrates? *In: Dead wood matters: the ecology and conservation of saproxylic invertebrates in Britain,* eds. by K.J. KIRBY & C.M. DRAKE, English Nature Science No. 7.

KEY, R.S. 1996. Invertebrate conservation and pollards. *In: Pollard and veteran tree management II;* ed. by H.J. READ, 21-28, Corporation of London.

KEY, R., & BALL, S.G. 1993. Positive management for saproxylic invertebrates. In: *Dead wood matters: the ecology and conservation of saproxylic invertebrates in Britain,* ed. by K.J. KIRBY & C.M. DRAKE, English Nature Science No. 7. Peterborough: English Nature.

KIRBY, K.J. 1988. *A woodland survey handbook.* NCC Research and survey in nature conservation. No. 11. Peterborough: Nature Conservancy Council.

KIRBY, K.J., REID, C.M., THOMAS, R.C., & GOLDSMITH, F.B. 1998. Preliminary estimates of fallen deadwood and standing dead trees in managed and unmanaged forests in Britain. *Journal of Applied Ecology*, **32**: 148-155.

KIRBY, P. 1992. *Habitat management for invertebrates: a practical handbook.* JNCC/RSPB.

KOZLOWSKI, T.T., KRAMER, P.J., & PALLARDY, S.G. 1991. *The physiological ecology of woody plants.* Academic Press Inc. San Diego.

LANE, A., & TAIT, J. 1990. *Practical conservation - Woodlands.* Hodder & Stoughton, Open University.

LEGG, R. 1995. Oaks at Windsor. *Tree News*, Autumn 1995:18.

VETERAN TREES
I N I T I A T I V E

LE SUEUR, A.D.C. 1931. Burnham Beeches: A study of pollards. *The Quarterly Journal of Forestry* **1931**: 12-25.

LE SUEUR, A.D.C. 1934. *The care and repair of ornamental trees in garden, parks and street.* London: Countrylife.

LEWIS, L., GOTHAM, P., OTTERBURN, B., OVERBURY, T., SHEPHERD, P., & BACON, J. 1997. Bracken breaking, a bruising battle. *Enact,* 5(3): 21.

LEWIS, N.R., & SHEPHERD, P.A. 1996. The management of bracken (*Pteridium aquilinum*) in the ancient Sherwood Forest, Nottinghamshire. *In: Pollard and veteran tree management II;* ed. by H.J. READ, 21-28, Corporation of London.

LONSDALE, D. 1991. Pollarding success or failure; some principles to consider. *In: Pollard and veteran tree management;* ed. by H.J. READ, 57-58, Corporation of London.

LONSDALE, D. 1993. *A comparison of 'target' pruning, versus flush cuts and stub pruning.* Arboriculture Research Note 116/93/PATH.

LONSDALE, D. 1994. *Choosing the time of year to prune trees.* Arboriculture Research Note No. 117.

LONSDALE, D. 1996. Pollarding success of failure; some principles to consider. *In: Pollard and veteran tree management II;* ed. by H.J. READ, 100-104, Corporation of London.

LONSDALE, D. 1999a. *The principles of tree hazard assessment and management.* Research for Amenity Trees <u>5</u>. TSO. London.

LONSDALE, D. 1999b. *Hazards from trees: a general guide.* Forestry Commission Practice Guide. Edinburgh: Forestry Commission.

MACMILLAN, P.C. 1988. Decomposition of coarse wood debris in an old growth Indiana forest. *Canadian Journal of Forest Research,* **18**: 1353-1362.

MARREN, P. 1990. *Woodland heritage.* David & Charles: Newton Abbot & London.

MARREN, P. 1992. *The wildwoods.* David & Charles: Newton Abbot & London.

MATTHECK, C., BETHGE, K., & Erb, D. 1993. Failure criteria for trees. *Arboriculture Journal,* **17**: 201-209.

MATTHECK, C., & BRELOER, H. 1994. *The body language of trees.* Research for amenity trees No. 4. London: HMSO.

MCLEAN, I.F.G., & SPEIGHT, M.C.D. 1993. Saproxylic invertebrates - The European context. *In: Dead wood matters: the ecology and conservation of saproxylic invertebrates in Britain,* eds. by K.J. KIRBY and C.M. DRAKE, English Nature Science No. 7.

MAYHEW, C. 1993. A new look at old tree practices. *Horticulture Week,* April 9 1993: 21-23.

MERCER, L. 1993. Pollard guidelines. Unpublished project for The National Trust (North West Region) and Houghall College.

MITCHELL, A.F. 1974. *A field guide to the trees of Britain and Northern Europe.* Collins.

MITCHELL, F.J.G., & KIRBY, K.J. 1990. The impact of large herbivores on the conservation of semi-natural woodlands in the British uplands. *Forestry,* **63**: 333-353.

MITCHELL, P.L. 1989. Repollarding large neglected pollards: A review of current practice and results. *Arboricultural Journal,* **13**: 125-142.

MITCHELL-JONES, A.J., & MCLEISH, A.P. (Eds.) 1999. *The bat worker's manual.* 2nd edition. Peterborough: JNCC.

ØRUM-LARSEN, A. 1990. The old north European 'meadow copse' and the English Landscape Park. *Garden History,* **18**(2): 174-179.

PATCH, D. 1991. Some thoughts on the physiology of pollarding. *In: Pollard and veteran tree management;* ed. by H.J. READ, 56, Corporation of London.

PATCH, D., COUTTS, M.P., & EVANS, J. 1986. *Control of epicormic shoots on amenity trees.* Arboricultural Research Notes 54/86/SILS

PAVIOUR-SMITH, K. & ELBOURN, C.A. 1993. A quantitative study of the fauna of small dead and dying wood in living trees in Wytham Woods, near Oxford. In: *Dead wood matters: the ecology and conservation of saproxylic invertebrates in Britain,* eds. by K.J. KIRBY, and C.M. DRAKE, English Nature Science No. 7.

PETERKEN, G.F. 1996. *Natural woodland.* Cambridge: Cambridge University Press.

PETERKEN, G.F., SPENCER, J.W., & FIELD, A.B. 1996. *Maintaining the Ancient and Ornamental Woodlands of the New Forest.* Consultation Document. Forestry Commission.

PHIBBS, J. 1991. Groves and belts. *Garden History,* **19**(2): 175-187.

PHILLIPS, J.B. 1971. Effect of cutting techniques on coppice re-growth. *Quarterly Journal of Forestry,* **65**: 220-223.

POTT, R. 1989. The effects of woodpasture on vegetation. *Plants Today,* September-October: 170-175.

QUELCH, P. 1996. Ancient trees in Scotland. *In: Pollard and veteran tree management II;* ed. by H.J. READ, 82-85, Corporation of London.

QUELCH, P.R. 1997. Ancient trees in Scotland. *In: Scottish woodland history,* ed. by T.C. SMOUT, Edinburgh: Scottish Cultural Press.

RACKHAM, O. 1986. *The history of the countryside.* London: J.M. Dent & Sons Ltd.

RACKHAM, O. 1988. Trees and woodland in a crowded landscape - The cultural landscape of the British Isles. *In: The cultural landscape - Past, present and future;* eds. by H.H. BIRKS, H.J. BIRKS, P.E. KALAND and D. MOE, 53-77, Cambridge University Press.

RACKHAM, O. 1989. *The last forest.* London: J.M. Dent & Sons Ltd.

RACKHAM, O. 1990. *Trees and woodland in the British landscape.* London: J.M. Dent & Sons Ltd.

RACKHAM, O. 1991. Introduction to pollards. *In: Pollard and veteran tree management;* ed. by H.J. READ, 6-10, Corporation of London.

RACKHAM, O. 1995. Bundle planting. *Tree News,* Autumn 1995.

RADLEY, J. 1961. Holly as a winter feed. *Agricultural History Review,* **9**: 89-92.

RAYNER, A.D.M. 1996. The tree as a fungal community. *In: Pollard and veteran tree management II;* ed. by H.J. READ, 6-9, Corporation of London.

READ, H.J. (ed.) 1991. *Pollard and veteran tree management.* Corporation of London

READ, H.J. (ed.) 1996. *Pollard and veteran tree management II.* Corporation of London

READ, H.J., FRATER, M., & Noble, D. 1996. A survey of the condition of the pollards at Burnham Beeches and results of some experiments in cutting them. *In: Pollard and veteran tree management II;* ed. by H.J. READ, 50-54, Corporation of London.

READ, H.J., FRATER, M., & TURNEY, I.S. 1991. Pollarding in Burnham Beeches, Bucks.: A historical review and notes on recent work. *In: Pollard and veteran tree management;* ed. by H.J. READ, 11-18, Corporation of London.

REED, P. 1996. The ecological value of tree management and its significance to species of epiphytic moss. *In: Pollard and veteran tree management II;* ed. by H.J. READ, 17-18, Corporation of London.

REID, C. 1996. Management of veteran trees on National Nature Reserves. *In: Pollard and veteran tree management II;* ed. by H.J. READ, 105-110, Corporation of London.

REID, C.M., FOGGO, A. & SPEIGHT, M. 1996. Dead wood in the Caledonian pine forest. *Forestry,* **69**(3): 275-279.

ROSE, F. 1976. Lichenological indicators of age and environmental continuity in woodlands. *In: Lichenology: Progress and problems.* eds. by D.H. BROWN, D.L. HAWKSWORTH and R.H. BAILEY; 279-307. London: Academic Press.

ROSE, F. 1991. The importance of old trees, including pollards, for lichen and bryophyte epiphytes. *In: Pollard and veteran tree management;* ed. by H.J. READ, 28-29, Corporation of London.

ROSE, F. 1993. Ancient British woodlands and their epiphytes. *British Wildlife,* **5**(2): 83-93.

RUSH, M.J. 1999. *Veteran Trees Initiative: Historical and cultural aspects a bibliography.* English Nature Research Report No. 318. Peterborough: English Nature.

SANDERSON, N. 1991. Notes on Holly cutting in the New Forest. *In: Pollard and veteran tree management;* ed. by H.J. READ, 53-55, Corporation of London.

SANDERSON, N. 1996a. The role of grazing in the ecology of lowland pasture woodlands with special reference to the New Forest. *In: Pollard and veteran tree management II:* ed. by H.J. READ, 111-117, Corporation of London.

SANDERSON, N.A. 1996b. *Lichen conservation within the New Forest timber inclosures.*

Vol. 1 Summary of survey and recommendations. Hampshire Wildlife Trust.

SANDERSON, N.A. 1998a. *Glen Finglass historic landscape survey, final report 1998.* A botanical survey & assessment report to The Woodland Trust.

SANDERSON, N.A. 1998b. Veteran trees in Highland pasture woodland. *In: Scottish Woodland History Discussion Group* Notes III. ed. by T.C. SMOUT, 4-11. Scottish Natural Heritage.

SANDERSON, N.A. in prep. Woodland management and lichens. *In: Habitat management and lichens.*

SCHEIDEGGER, C., FREY, B., & ZOLLER, S. 1995. Transplantation of symbiotic propagules and thallus fragments: Methods for the conservation of epiphytic lichen populations. *In: Mitt. Eidgeöss. Forsch. Anst. Wald Schnee Landsch,* eds. by C. SCHEIDEGGER, P.A. WOLSELEY and G. THOR, **70**: 41-62.

SEARLE, S.H. 1996. Management of veteran trees in Windsor Forest. *In: Pollard and veteran tree management II;* ed. by H.J. READ, 67-68, Corporation of London.

SHIGO, A.L. 1986a. *A new tree biology.* Shigo and trees associates. Durham: USA.

SHIGO, A.L. 1986b. Journey to the center of a tree. *American Forests,* June 1986: 1-6.

SIDWELL, R.C.G. 1996. Hainault Forest Country Park: Re-pollarding hornbeam trials. *In: Pollard and veteran tree management II;* ed. by H.J. READ, 65-66, Corporation of London.

SISITKA, L. 1991a. Pollarding experiences at Hatfield Forest, Essex. *In: Pollard and veteran tree management;* ed. by H.J. READ, 19-21, Corporation of London.

SISITKA, L. 1991b. Pollarding. The Hatfield Forest Experience. *Tree News,* September 1991: 15-16.

SMART, N., & ANDREWS, J. 1985. *Birds and broadleaves handbook.* RSPB.

SMOUT, C., & Watson, F. 1997. Exploiting semi-natural woods, 1600-1800. *In: Scottish woodland history,* ed. by T.C. Smout: 86-100.

SMITH, D. 1991. The management of the oaks at Kingston Lacy Estate, Wimborne, Dorset. *In: Pollard and veteran tree management;* ed. by H.J. READ, 22, Corporation of London.

SPEIGHT, M.C.D. 1989. *Saproxylic invertebrates and their conservation.* Nature and Environment Series No. 42. Council of Europe, Strasburg.

SPENCER, J., & FEEST, A. (Eds.) 1994. *The rehabilitation of storm damaged woods.* University of Bristol.

SPRAY, M. 1981. Holly as fodder in England. *Agricultural History Review,* **29**: 97-110.

TUBBS, C.R. 1986. *The New Forest.* London: Collins, New Naturalist.

TUBBS, C.R. 1997. The ecology of pastoralism in the New Forest. *British Wildlife,* **9**(1): 7-16.

TUSSER, T. 1573, 1577 & 1580. *Five hundred pointes of good husbandrie.* London.

VETERAN TREES
INITIATIVE

U.K. BIODIVERSITY STEERING GROUP 1998. Lowland wood pasture and parkland: A habitat action plan. *In:* Tranche *2 Action Plans Vol. II. Terrestrial and freshwater habitats.* Peterborough: UKBG/English Nature.

WALL, T. 1991. Managing veteran holly trees - A preliminary note. *In: Pollard and veteran tree management;* ed. by H.J. READ, 51-52, Corporation of London.

WALL, T. 1996. Strategies for nature conservation in parklands: Some examples from Moccas Park National Nature Reserve. *In: Pollard and veteran tree management II;* ed. by H.J. READ, 42-49, Corporation of London.

WARRINGTON, S., & BROOKES, R.C. 1998. The recovery of hornbeam *Carpinus betulus* following the reinstatement of pollard management. *For. & Landsc. Res. 1998* (**1**): 521-529.

WATKINS, C. 1990. *Woodland management and conservation.* Newton Abbot & London: David & Charles.

WATKINS, C., & GRIFFIN, N. 1993. The liability of owners and occupiers of land with large old trees in England and Wales. *In: Dead wood matters: the ecology and conservation of saproxylic invertebrates in Britain,* eds. by K.J. KIRBY and C.M. DRAKE, English Nature Science No. 7.

WATSON, F. 1997. Rights and responsibilities: wood-management as seen through baron court records. *In: Scottish Woodland History,* ed. by T.C. SMOUT: 101-114.

WHEAL, A. 1998. Some pollarding techniques can do more harm than good. *Horticulture Week,* March 12: 23-24.

WHITE, J. 1991a. Suggestions for re-pollarding oaks at Markshall, Essex. *In: Pollard and veteran tree management;* ed. by H.J. READ, 46-47, Corporation of London.

WHITE, J. 1991b. Dating the veterans. *Tree News,* Spring/Summer 1995: 10-11.

WHITE, J. 1996. Progress with re-pollarding old oaks and new work on ash. *In: Pollard and veteran tree management II;* ed. by H.J. READ, 75-76, Corporation of London.

WHITE, J. 1998. Estimating the age of large and veteran trees in Britain. Forest Information Note 250.

WIGNALl, T.A., BROWNING, G., & MACKENZIES, K.A.D. 1987. The physiology of epicormic bud emergence in pedunculate oak (*Quercus robur* L.) responses to partial notch girdling in thinned and unthinned stands. *Forestry,* **60**(1): 45-56.

WINTER, T. 1993. Dead wood - Is it a threat to commercial forestry? *In: Dead wood matters: the ecology and conservation of saproxylic invertebrates in Britain,* eds. by K.J. KIRBY & C.M. DRAKE, English Nature Science No. 7.

WISDOM, K. 1991. Pollarding experiences of the Woodland Trust. *In: Pollard and veteran tree management;* ed. by H.J. READ, 50, Corporation of London.

INDEX: VETERAN TREES

FIGURES AND ILLUSTRATIONS ARE IN *ITALICS*.

VETERAN TREES
INITIATIVE